Model-driven Development for Embedded Software

Model-driven Development for Embedded Software

Application to Communications for Drone Swarm

Jean-Aimé Maxa
Mohamed Slim Ben Mahmoud
Nicolas Larrieu

ISTE PRESS

ELSEVIER

First published 2018 in Great Britain and the United States by ISTE Press Ltd and Elsevier Ltd

ISTE Press Ltd
27-37 St George's Road
London SW19 4EU
UK

www.iste.co.uk

Elsevier Ltd
The Boulevard, Langford Lane
Kidlington, Oxford, OX5 1GB
UK

www.elsevier.com

British Library Cataloguing-in-Publication Data
A CIP record for this book is available from the British Library
Library of Congress Cataloging in Publication Data
A catalog record for this book is available from the Library of Congress
ISBN 978-1-78548-263-2

Contents

Preface

The aim of this book is to describe the principles of model-oriented design used in the field of aeronautics, specifically for unmanned aerial vehicles (UAVs) or drones.

In this book, we will focus on the design of an on-board system for UAV ad hoc communications. In this context, we present an original rapid prototyping methodology for complex embedded systems, showing how this approach creates considerable time savings in the verification and formal validation phases, contributing to UAS (Unmanned Aerial System) certification.

We will also discuss more traditional, but necessary, verification phases which must be carried out in order to verify system performances. This evaluation is conducted through network simulation and testbed experimentations.

The different tools required to implement this methodology will also be described in order to allow readers to reproduce all or part of the approach themselves.

Finally, in order to illustrate the benefits of our new approach, we provide an example of use through the development of an embedded system in the field of aeronautics, specifying the different phases of the methodology. The aim is to design, validate and test a new secure routing protocol for UAV ad hoc communications.

Jean-Aimé MAXA
Mohamed Slim BEN MAHMOUD
Nicolas LARRIEU
January 2018

Introduction and Approach

The drone industry is rapidly evolving. The type and the usage of industrial drones have changed considerably over the last five years. Drones are now able to carry increasingly complex payloads, with unprecedented levels of autonomy and automation during their assigned missions.

This increase in UAV complexity levels requires improvements in the processes and methods used for their design and evaluation and for the success of missions in which they are involved. The aim of this book is to present a new rapid prototyping method, intended for the design of complex embedded systems using simple and intuitive design tools. The work presented here is inspired by previous contributions to the aeronautical domain, where the complexity inherent in the development of embedded systems has received considerable attention over the past few decades.

The work presented in this book is innovative in terms of the relevance of the rapid prototyping method presented in Chapter 2, and also in terms of the application of this method. The communicating drone network project which will be presented later in this book is one of the very first experiments in which multiple UAVs, with shared mission objectives, have been able to exchange surveillance information (video) securely and in real time. Thus, in this implementation, security is dependent on the type of communication network (an ad hoc network in which each drone may act as an emitter, relayer or receiver of information), and also on the security mechanisms applied to information exchanges during the fleet mission. Note that all of the protocols presented later in this work were defined, designed and evaluated using the rapid prototyping method presented here. To the best of our knowledge, no other similar work in the field of embedded systems has involved the application of model-oriented methods to the specific context of communicating drone networks.

The rest of this work will be organized as follows. Chapter 2 is given over to a state of the art of model-driven development methods applied to aeronautical systems. Drones are usually considered as autonomous aircraft, as the software requirements are similar to those for conventional aircraft. It is thus interesting to compare existing approaches to MDD (Model-Driven Development) for traditional aircraft (e.g. an Airbus A380) with those used for UAS (Unmanned Aerial Systems).

In Chapter 2, we will present our prototyping method for embedded drone systems. This original method is built on MDD principles in order to design complex systems (e.g. a communicating drone network) with the assistance of top-level artifacts. These artifacts rely on the use of a model-driven formalism, allowing simple and rapid definition of the final system functions. These high-level models have a higher power of expression than classic software specifications, and thus simplify the validation of system functionalities. Moreover, the use of high-level models creates new possibilities in terms of formal verification methods, which will also be discussed in this chapter. The phase in which the functionalities of the final system are validated and verified is critical for aeronautical systems (including drones); the certification requirements for flight authorization are particularly stringent in this case. Formal verification methods, associated with the use of high-level models for system design, make it possible to reduce the engineering workload involved in the software validation phase which follows modeling. These different points will be discussed in detail in Chapter 2, along with a discussion of the advantages obtained by using formal methods in conjunction with high-level models. Note, however, that formal validation of high-level models is not sufficient to verify all functionalities of the final system. This first phase of formal verification must be followed by a more traditional verification phase (e.g. through unitary testing). Our discussion will therefore also cover more "standard" verification tools used to validate functionalities of the final system. We will pay particular attention to a hybrid simulation tool developed specifically for the purposes of validating network operations (at protocol level) for drone fleets. We will also focus on the physical components required to implement information exchange functions in the embedded system in question (i.e. the Delair-Tech DT 18 drone).

Finally, we will give a detailed account of the implementation which led us to define the rapid prototyping method presented in the previous chapter. This implementation exemplifies a very promising application of drone fleets, involving the simultaneous use of multiple UAVs to cover a far larger geographical area than would be possible with one drone. For drones to operate as a fleet, they need to be able to communicate in order to reduce the number of control and information exchange stations needed for the surveillance mission. This principle leads to the definition of an ad hoc drone network. In this context, new communication protocols were defined and implemented using our rapid prototyping method. The final chapter of this work notably includes the presentation of a new routing protocol, which takes account of the vulnerabilities inherent in ad hoc communication networks, and proposes new mechanisms for efficiently solving these issues. The improvements to the routing protocol are intended to increase the security of the ad hoc network, improving service for surveillance missions. The final chapter will describe the model-driven development process for the new routing protocol. This evaluation will focus on three aspects: the use of formal methods; the use of a hybrid simulation tool; and real experiments, in which multiple drones were involved in a geographically distributed surveillance mission.

1

State of the Art of Model-driven Development (MDD) as Applied to Aeronautical Systems

1.1. Principle of MDD

Faced with an exponential increase in program complexity, operators in the aeronautical sector have established software-based certification procedures based on the use of model-driven methods. These methods guarantee a certain level of operational security, and in some cases make the design process easier.

Generally speaking, software which is embedded in a critical system such as an airplane or other aircraft must be subject to certain certification constraints in order to be considered trustworthy. Certification implies a certain degree of confidence in the system. For software in particular, it is important to show that the design follows a development process in accordance with the state of the art in the aeronautical sector.

Most software design methods are based on UML (Unified Modeling Language) [RUM 04]. However, these methods need to be adapted to take account of the operating environment of the final system. Methods based on UML only allow high-level descriptions of a system, with no consideration for the constraints involved in its physical implementation and execution. Moreover, UML does not respond to design requirements in the aeronautical context, or in the case of drones. This is due to the fact that it does not possess the toolchains required to contribute to the validation of a critical system. In the context of designing an embedded software program for manned or unmanned aircraft, software certification must be taken into account during the design phase. This consideration implies the use of chains of design tools which contribute to the attainment of certification for the final system.

Model-driven approaches aim to generate some or all of a system through the use of high-level models. This paradigm increases productivity while also optimizing compatibility between different sub-systems, thanks to widespread reuse of normalized models. This also simplifies the software design process and facilitates model reuse due to the levels of abstraction encountered in the associated professional logic.

Model-driven approaches use models to improve the forecasting, design, implementation and modification of systems. They offer a number of advantages. First, they encourage the efficient use of high-level models in the design process. They also offer the possibility of using better design practices in system creation. The main aims of the MDD approach include portability, interoperability and reusability, via the separation of platform-dependent aspects and more abstract aspects which are not dependent on a specific application. This type of approach was introduced and defined by the OMG (Object Management Group), which aimed to develop the object-oriented approach while increasing the level of abstraction to the point of using another representation of concepts and relationships drawn from an initial specification, i.e. the model. A model is an abstract representation of the knowledge and activities which govern a domain of application, making it easier to understand the final system. This development technique allows designers to focus on desired system behaviors rather than on implementation. The partial generation of code using model specifications leads, among other things, to savings in terms of development costs.

1.2. Use in avionics

Improvements in the performance of aeronautical systems mean that it is now possible to envisage the use of new technologies in the context of embedded aeronautical systems on aircraft, along with the opening up of avionic networks, previously closed for security reasons, to public networks such as the Internet. These new technologies require new solutions in order to maintain the high levels of security required.

1.2.1. *System virtualization: Integrated Modular Avionics*

The first generations of avionic software systems were based on direct relationships between systems: when a captor transmitted an element of information to two on-board computers, the data was duplicated and sent over two independent communication channels, each serving a single receiver. The development of new technologies has resulted in the creation of new services for crews and in the introduction of new interactions.

A new concept, Integrated Modular Avionics (IMA), was introduced with the development of the A380. It allows several independent programs to be executed within a single hardware module. RTCA (Radio Technical Commission for Aeronautics) DO-297, the Integrated Modular Avionics Development Guidance and Certification Considerations standard of 8th November 2005, sets out a framework for the design and implementation of systems for integrated modular avionic architectures in civil aviation. Created by Special Committee 200 (SC-200), this standard defines IMA as "a shared set of flexible, reusable, and interoperable hardware and software resources that, when integrated, form a platform that provides services, designed and verified to a defined set of requirements, to host applications performing aircraft functions". This standard defines and delimits the roles of different IMA module suppliers: application suppliers, IMA platform suppliers, system integrators and certification agents.

1.2.2. *MILS: divide and conquer to ensure security*

The segregation of participants which occurs in virtualization solutions makes it useful for increasing the security of sensitive applications. This observation led to the gradual development of the concept of Multiple [and] Independent Levels of Security[/Safety] (MILS) architectures. Based on John Rushby's work on micro-kernel separation, MILS architectures guarantee a high level of security for the execution of multiple programs in a single infrastructure.

The division of a complex task into several simpler tasks is a fundamental principle of engineering. In computing, this equates to decomposing or dividing a program into modules. This simplifies security assessments, as the evaluator does not need to evaluate a whole, monolithic system, but rather a set of smaller, distinct modules and pairings.

Virtualization solutions may be implemented for developing a support for MILS architecture on the condition that they guarantee respect for the four intrinsic properties of MILS:

– the solution must be impossible to circumvent, i.e. no entity may communicate with the system without passing the security checks imposed by the host system;

– it must be possible to evaluate the solution, with formal proof that the virtualization system (and thus the host) operates in a correct and valid fashion;

– the solution must always be active: all communications are monitored, not just the first messages exchanged;

– the solution must be resistant to alteration, preventing any modification without explicit authorization.

These properties are guaranteed via an evaluation of the security of the solution. Even for "small" systems, this evaluation is complex; it is only achievable for minimalist systems (microsystems) intended for system virtualization and separation, known as separation microkernels. Separation microkernels ensure the implementation of concepts of temporal and spatial separation between programs, while monitoring information flows. The kernel ensures that each program and its virtual machine (VM) are able to use hardware resources during their assigned times. A program cannot hinder the operation of another program, "stealing" its operating time; there is thus a temporal separation between the two VMs. The kernel also guarantees that a hardware resource will not be simultaneously assigned to two virtual machines. The addressing spaces in the memory and input/output channels are shared out during configuration of the separation kernel: this is known as partitioning the addressing space.

Each instance of execution at each access point is checked by the kernel to ensure that the address accessed by the real machine is that which has been assigned to it. The kernel will block any access attempts which do not fulfill this condition, ensuring the spatial separation of VMs. Similarly, VMs may make use of specific channels to communicate with each other, which are also managed by the separation kernel (rather than operating directly through the subjacent electronics). The kernel thus monitors the form of communications, checking the maximum length of sent messages, authorizing access, stamping received messages, etc. In an MILS system, each virtual machine operates independently of its counterparts. From the perspective of the host separation kernel, each VM has its own dedicated addressing sub-space. The set of these sub-spaces is a partition (in the mathematical sense of the term) of the host's full addressing space. Each VM has specific assigned time slots within a cycle, and these slots make up a time partition. In the context of an MILS architecture, the term "partition" therefore relates to the temporal and spatial resources associated with a virtual machine. During system execution, the separation kernel acts as a partition scheduler and as a compulsory point of passage to access resources.

An MILS system needs to guarantee certain properties:

– in terms of inter-partition information flows, only authorized sources should be able to generate information, and information should only be delivered to specified, authorized recipients;

– data in a partition should only be accessible to programs associated with the partition in question, and should be isolated from data from other partitions. Private data remains private, with no possibility of infiltration (reading from another partition) or exfiltration (writing data to another partition). This means that each partition must have a dedicated addressing space, in which addresses only have meaning for the partition which uses that space;

– the processor itself must not allow information to travel from one partition to another, whether through material caches or even in measures of processing time. For example, a form of attack exists which involves analyzing processing cycle counters to extract information relating to cryptographic keys, as discussed in (Kocher, 1996). MILS systems must be as resistant as possible to attacks of this kind;

– operating errors in one partition should not have an impact on the other partitions, and should be detected, contained and corrected. MILS systems, used in military and civil information systems, have attracted growing interest in the aeronautical sector. However, virtualization is already used in this area, with a slightly different goal - that of enabling hardware sharing while maintaining high levels of security (operational safety), as we saw in the case of IMA. The current aim in aeronautics is to allow these two aspects (security and safety) to be treated together, as we will see in the following section.

1.2.3. *Combined treatment of safety and security considerations*

The development of embedded systems in aircraft is subject to significant constraints, both in terms of operational safety and security. In aeronautics, the term “safety” is used to denote both the security and the operational safety of systems, i.e. the intrinsic properties of systems which make them resistant to operating errors. The term “security” relates to the ability of aeronautical systems to resist deliberate attacks (pirating, etc.).

The safety constraints encountered in the domain of avionics have always been particularly stringent. Any general operating fault in an aircraft may lead to its partial or total destruction, endangering human lives. Different aeronautics standards specify safety constraints which must be respected in order for aircraft to be allowed to fly. Critical embedded avionic programs, such as autopilots, are subject to standard RTCA DO-178B.

DO-178B defines five Design Assurance Levels (DAL) for the development of avionic software. DAL-A is the highest and most restrictive level and is used for critical applications where an operating fault could have catastrophic consequences, while DAL-E is much less restrictive, and is reserved for applications which do not affect the safety of the aircraft in any way. The standard DO-178B and the different levels of DAL will be discussed in greater detail in section 1.3.2. Programs must be evaluated on the basis of safety criteria before being embedded and used in an aircraft. This verification activity, known as certification, is carried out by independent organizations at the national level. In France, for example, aeronautical systems are certified by the DGAC (Direction Générale de l’Aviation Civile).

Originally, aerospace firms focused their in-depth work on safety considerations, considering the complexity of embedded systems and the closure of avionic networks to be sufficient to guarantee security. Since the 9/11 attacks, however, security has become a major consideration in system design, which has led to the introduction of new constraints and new practices. Sensitive products must be security checked via an evaluation process. Security requirements may be grouped into “packets” of requirements for the purposes of evaluating the final products. There are seven of these packets for assurance requirements alone, known as the Evaluation Assurance Levels or the EAL, and number from 1 to 7. EAL-2, for example, encompasses all of the requirements of EAL-1 plus a number of additional elements, and so on up to EAL-7, the highest level of security assurance which may be assigned according to the CC ITSEC (Common Criteria for Information Technology Security Evaluation) nomenclature. This standard has been adopted by the ISO as ISO CEI 15408, and is often referred to simply as the "Common Criteria" or the CC; it is used to evaluate whole systems with regard to security requirements.

Despite the existence of these two standards (DAL and EAL), certification (guaranteeing system safety) and evaluation (guaranteeing system security) are currently treated independently and in parallel; certain very similar verifications are thus carried out twice. Work on defining a new development method covering both safety and security aspects, while avoiding redundancy in the steps and processes involved, is currently underway. However, this lies outside the scope of our work here, and will not be addressed further.

1.2.4. *Certification of an avionics system*

1.2.4.1. *Qualification of tools for certification*

The cost of certifying avionic software, in terms of time, man-hours and money, increases as the required level of assurance increases. An airplane autopilot requiring DAL-A certification can take hundreds of engineers several years to define, at a cost reaching into the millions of dollars; DAL-A certification involves 66 independent checks. The program controlling "No Smoking" and "Fasten Seat Belt" light displays in aircraft, however, only requires DAL-D certification, involving a mere 15 control tasks. The development of this program is much less costly due to its low impact on aircraft safety. During the design process, certification costs may be reduced by automating certain control tasks via the use of specific tools which guarantee certain safety properties. The use of a DAL-A qualified compiler for conformity, for example, guarantees that the binary code conforms to the source code, fulfilling the "verify the conformity of binary code with source code" requirement for DAL-A certification. However, the tool (in this case, the compiler) itself needs to be verified to ensure that it does not introduce errors. The verification of software tools is known as qualification, and is carried out in a similar way to aeronautical software certification.

1.2.4.2. *Model-driven design approaches in aeronautics*

MDD (Model-driven Development) approaches have long been used in traditional software engineering. Unified Modeling Language (UML) is the most widespread language used across all industries as a support for design methods when defining and implementing classic software. However, this toolkit cannot be used for certain specific developments, including embedded software in the aeronautic and aerospace industries. The main issue with UML-based design methods lies in the fact that the tools used to produce final software systems are unable to take account of certification requirements. This is not necessarily problematic for mass market applications (e.g. when designing web services) which may make use of UML, or for industrial applications with limited constraints. However, it must be taken into account in the case of critical applications, such as aircraft or satellites. Certification of the final product must be taken into account, and the toolchains used in the appropriate standardization process for applications with high security and safety requirements must be integrated.

Note that the certification process for embedded software designed for specific uses in the aeronautical industry is highly and precisely codified. There are several standards which must be taken into account. Here, we will only consider those relating to software engineering for embedded systems: DO-178 C and DO-331.

1.2.4.3. *DO-178C: Software Considerations in Airborne Systems and Equipment Certification*

This document was issued by the RTCA in 2012 and constitutes the fourth edition of a document that defines the applicable standards for certifying embedded avionic software systems. This latest version takes account of the latest developments in MDD for software engineering. Notably, it takes account of the possibility of validating systems using formal verification methods for the first time in an aeronautical context; this approach can reduce the number of unitary tests required for validation of a final product. It constitutes a significant advance in aeronautical engineering, with a profound impact on the way in which aeronautical systems will be designed and produced in future. The document highlights MDD approaches which are able to automatically generate source code from high-level models, defining the functionalities and behaviors of the final system.

1.2.4.4. *DO-331: Model-Based Development and Verification*

DO-331 is a supplement to the DO-178C standardization document which suggests classes of tools and methods which may be used to automatically generate source code and to validate the high-level models used in the automatic code generation process. DO-331 recommends the use of model-based verification methods. The RTCA notably recommends three types of mechanisms to facilitate the verification process: model checking, formal proofs and code assertions. These different verification techniques have been used for several years in other areas of industry, and have now reached a sufficient level of maturity for application in complex environments, such as aeronautical validation and certification processes.

1.3. The case of drones (UAS - Unmanned Aerial Systems)

In the field of drones, the different perspectives for application offered by UAS show promise in terms of the possibility of carrying out flexible and evolutive missions. These missions are generally carried out over inhabited areas within national airspace. It is therefore essential to ensure that missions are carried out successfully, covering every eventuality for every system module, in order to prevent operational failures with potentially catastrophic consequences (e.g. loss of life). As yet, there is no operational safety standard for unmanned aircraft. A document of this type would be helpful in defining the lifecycle and security levels of the embedded systems making up the UAS. We will focus on this type of problem here, considering the engineering aspect of developing complex systems and proposing a rapid prototyping methodology for the development of a secure routing protocol for a fleet of drones.

1.3.1. *The need for a new rapid prototyping methodology for UAS design*

A drone system is generally made up of several software modules, each responsible for a set of specific functions. These modules are involved in dynamic exchanges of information relating to their environment, allowing them to offer different types of services during a given mission. To take account of this critical environment, operational safety assurances are needed when developing software for use within the UAS, ensuring maximum conformity between the specification and implementation of the source code. Note that in critical embedded systems, we speak of operational safety in terms of something which is dependent on the system operating correctly in response to input; the term does not cover information security aspects. The term is used here as it is widely used in existing literature on design studies for critical software.

Moreover, civil drones currently operate in specific airspaces, separate from the space used by short-, medium- and long-haul aviation. However, UAVs may need to share airspace with civil aircraft to fulfill certain commercial applications. For this to be possible, operational validation of the autopilot system will be required. This process consists of going through each of the systems involved in the UAS, studying their execution in every possible circumstance. Our contribution to the validation process is based on a solution adopted in the aeronautical field, making use of a rapid prototyping methodology in order to verify the operational safety of software during the design process. This method is based on a model-driven architecture, using a chain of formal verification and code generation tools. It ensures full traceability of requirements from the specification to their implementation in the source code. This method also offers portability and allows for the reuse of source code via the separation of platform-specific aspects and abstract aspects used to describe the system.

Our aim in this book is to illustrate the use of our model-driven rapid prototyping method via the implementation of a secure communication architecture in a drone fleet. This architecture is intended to supplement the software architecture which is already used in the Delair-Tech UAS[1]. While the current architecture already allows single drones to be flown in French civil airspace [MAN 15], the new system, involving a fleet of UAVs, needs to pass a further validation test. The use of drones in fleets results in new behaviors (particularly in cases where traffic command and control messages are transmitted through the fleet) which are not present in current sub-systems, meaning that a new global validation is required.

Our focus here is on contributing to validating our secure routing protocol through the use of a model-driven methodology. To validate this protocol, the

1 We worked in partnership with Delair-Tech on the implementation and real-world validation activities described in this book.

environment of use needs to be taken into account in order to create a final system which is as close as possible to the initial specification. By using models, the complexity of programs used in the drone system can be described in a clear and concise manner over several levels of abstraction. The use of models also allows us to create a modular system which is both easy to use and is reusable. The models in question are then translated into source code using the same code generation tools encountered in aeronautics. The models and code are then verified using formal verification tools in order to demonstrate the correspondence between the specification (using models) and the implementation of the functionalities described within the high-level models.

The use of verification and validation approaches is justified by the particular design requirements of critical embedded systems. The combination of specification constraints and safety constraints (at operational level) raises a number of questions; the use of automatic tools may offer a solution. In this context, we will now consider the influence of standards on embedded software design.

1.3.2. *Safety standards*

The safety standards governing the design of critical embedded systems are based on a lifecycle and a security level[2]. The lifecycle relates to different stages in the existence of embedded software, including operational safety aspects of the design process. The security level is used to assign requirements to each sub-system in relation to their criticality.

In this section, we will discuss standard DO-178B. This standard stipulates the methods which should be used to design and validate critical on-board systems in the field of avionics. It defines a set of rules, techniques and methods to apply for each desired security level. For example, DO-178B makes use of a five-level classification of failures defined by the FAA (Federal Aviation Administration). These levels of criticality are shown in Table 1.1. Level A, covering catastrophic failures which may prevent a mission from continuing,

2 Security concerns all of the human, organizational and technical resources involved in responding to spontaneous or premeditated acts with the intention to cause damage or prejudice to the system, in order to generate psychological or financial effects.

is the highest, while level E is used for parts of the system with no particular security requirements. Note that the level assigned to a system is determined by the risk assessment authority. Each level corresponds to a set of design rules which must be followed in order to obtain certification for embedded software. The operational safety recommendations from this standard may be applied to the domain of UAS in order to increase certification authority confidence in the final system.

Failure category	**Condition**
Level A	Catastrophic: failure conditions which would prevent continued safe flight and landing.
Level B	Hazardous: failure conditions which would reduce the capability of the aircraft.
Level C	Major: failure conditions which would reduce the capability of the aircraft, resulting in a significant reduction in safety margins or functional capabilities.
Level D	Minor: failure conditions which would increase crew workload while remaining within their field of capabilities; for example, a slight reduction in security margins or functional capabilities.
Level E	No safety effect: failure conditions with no effect on the operational capability of the aircraft.

Table 1.1. *DO-178B's failure classification*

1.3.3. *Software development lifecycle*

The development of critical embedded software involves a set of processes throughout its lifecycle. Each process includes one or more transition criteria which must be fulfilled in order to move on to the next stage. There are several software development cycles in existence [PON 08], but generally speaking, the cycle recommended in the applicable standard should be used. In the context of our study, there are currently no standards governing the design of embedded software for UAS. However, the field of aeronautics, which has broadly similar functional safety requirements [BRA 12], is governed by standard DO-178B which uses the V model. We have followed the recommendations set out in this standard, applying the V model as shown in Figure 1.1.

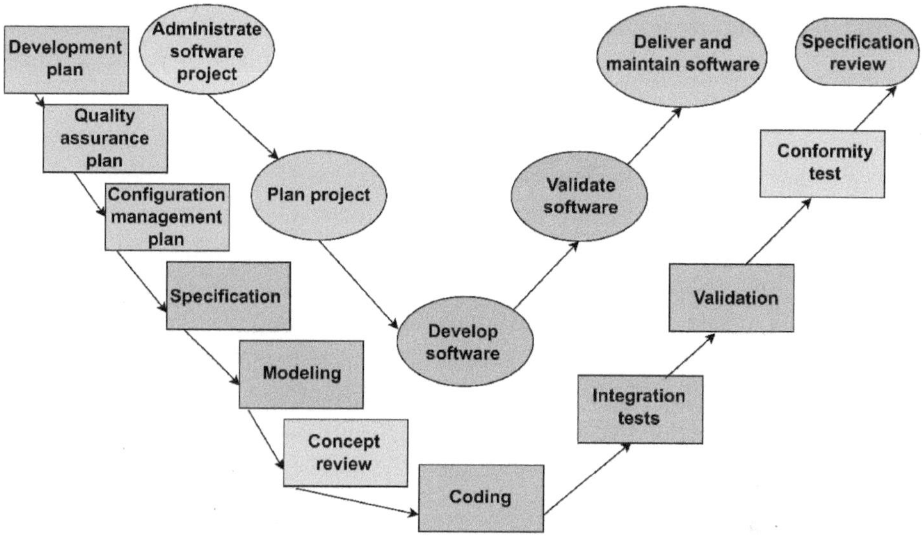

Figure 1.1. *The V development lifecycle. For a color version of this figure, see www.iste.co.uk/maxa/software.zip*

The V lifecycle process for software following DO-178B standards is generally made up of the following steps:

– the planning stage, in which development activities and project processes are defined and planned;

– program development, including specification, design, coding and integration phases;

– verification, which aims to detect requirement errors which may have arisen when defining the specification;

– configuration management, in which the configuration of the program is identified and traceability references are created. This stage also involves the creation of anomaly reports;

– quality assurance, with the aim of guaranteeing that the lifecycle produces a program which conforms to the specifications. This stage also includes a conformity review of the final product;

– coordination for certification, intended to provide the means of attaining program conformity and proof of the design plan. Data prepared and submitted to the certification authority includes plans of the software aspects of certification, the program configuration file and a summary of work that has been carried out.

Note that during the software development phase, software specifications make use of products from the system lifecycle to develop high-level requirements for the program. This includes functional, performance, interfacing and operational safety requirements. Finally, these specifications are modeled over several iterations. The addition of low-level code characterizing interaction with the operating system is traditionally carried out at this point (see [BAC 05] for details). The software is then coded by combining the generated code with additional glue code. It is then possible to move onto the software integration stage, in which executable code is loaded into the target machine.

In addition, DO-178 standards suggest techniques to use which conform to the desired safety level. For example, it recommends the use of a certified compiler (e.g. the Ada compiler[3]), or a compiler where confidence is attained through use (such as the GCC compiler[4]). These compilers contribute to validating the source code for any certification category (A, B, C, D or E) through their technical specificities.

Certification authorities may also prescribe the use of highly typed languages or coding rules. This is due to the fact that the constraints associated with the development of critical embedded software mean that programs belong to very specific categories. For example, the use of the *goto* command in critical programs is often strongly discouraged, as it is seen to be too difficult to study and to verify [DIJ 68]. Many other rules of this type exist for all levels of certification.

3 http://adacore.com.

4 http://gcc.gnu.org.

2

Original Rapid Prototyping Method for Embedded Systems for UAVs

2.1. Using models to auto-generate a system

Design tools such as MATLAB/Simulink® [NIC 09] or Lustre/SCADE [LAK 09] can help us to understand the complexity of a system by offering high levels of abstraction. They also allow formal methods to be used in the embedded software development lifecycle. These tools make it possible to model the environment in which the software is to be implemented. The general concept involves building the critical embedded program directly from models in order to limit design errors and to save time in the development process. While UML is often used for software modeling, MATLAB/Simulink presents a significant advantage, in that it enables a global view of the software system and its environment which contains enough information to validate behavior in relation to a specification. This tool is often used in industry for designing and testing avionic software [RIE 13]. Feedback has shown that MATLAB/Simulink is also useful in the certification process [EST 13].

In the next section, we will list the different stages in our methodology, based on the recommendations in standard DO-178B.

2.1.1. *Presentation of different steps*

Our proposed methodology is based on an adaptation of the V development lifecycle. We make simultaneous use of a model-to-code transformer, to shorten the coding phase, and a set of formal verification tools, used to verify the properties of the models and code. In this presentation, we will presume that the protocol specification stage has been completed prior to applying the methodology. We assume access to a set of clear and valid requirements defining all of the services which the final product must be able to offer (in this case, and in what follows, this product is a secure routing protocol), along with dependency relationships which must be taken into account. In the case of our secure routing protocol, this set of requirements notably includes elements relating to the routing of exchanged traffic, the security of these exchanges, and the performance of the final system.

Our methodology may be summarized as a set of seven successive steps, as seen in Figure 2.1.

The first step in our methodology is partitioning, which consists of dividing the future program into different partition classes. Each partition is associated with a sub-set of system functionalities (and thus with a sub-set of functional requirements). This operation is carried out manually and has a major impact on the rest of the design process. The purpose of the partitioning stage is to divide the specification into a group of distinct procedure sub-sets. The final program must execute a set of instances of these procedures.

The second step is devoted to modeling, describing the behavior of each software partition in the form of Simulink blocks and state machines. These models are easy to read and allow us to test and validate program operations at an early stage in the design phase. In this specific case, the routing protocol was modeled using Simulink and Stateflow®, version R2014B [GUI 14].

The third step in the methodology involves transforming each partition class model into source code using an automatic code generator. This is the transformation stage. Code may be generated in any language supported by the generator. In our case, we used the Embedded Coder tool [ROS 09] to generate code in C. Embedded Coder is a proprietary program produced by MathWorks.

During the fourth step, handwritten elements are added to the code generated previously to allow it to operate within the final system. These new elements are known as glue code. Note that glue code may also be created automatically in the case of more highly automated software production

[VAR 12]. This additional code contains the specification for connections between input/output in the model and in the software partition of the operating system in the target hardware. It also specifies sequencing information groups for the different partitions.

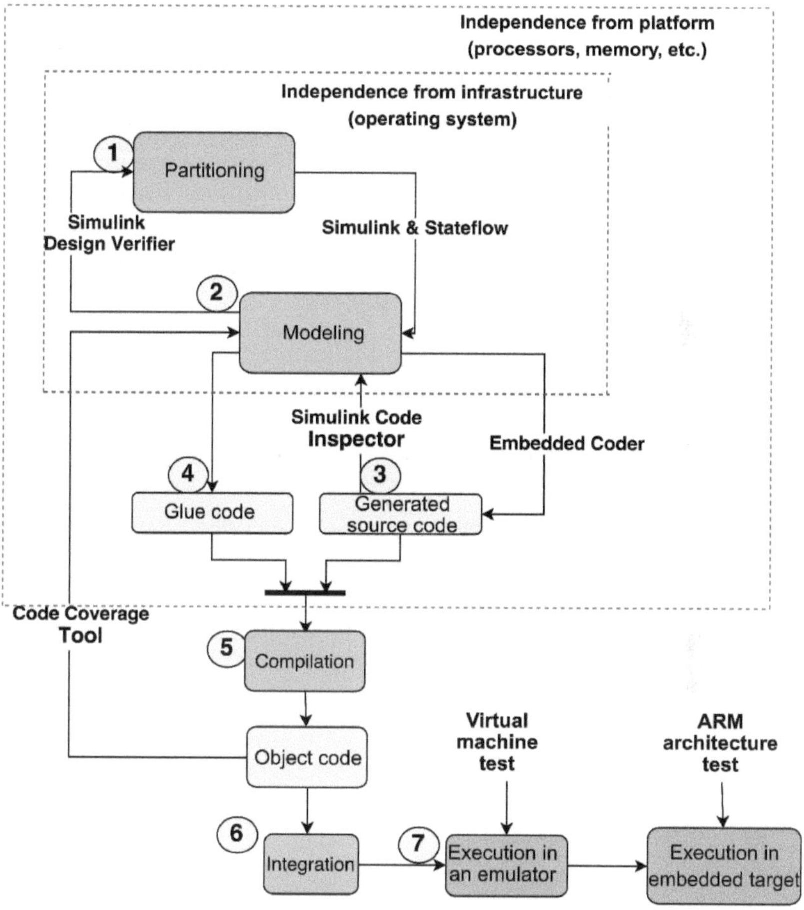

Figure 2.1. *Model-driven development methodology for embedded complex systems. For a color version of this figure, see www.iste.co.uk/maxa/software.zip*

The fifth step is compilation, whereby the automatically generated and handwritten code elements are compiled together to create binary code for all partition classes. This leads on to the sixth step, description of the global structure of the algorithm, including a specification of the number of instances executed in parallel for each partition, along with the associated configuration and sequencing information. These parameters are combined with the operating system kernel and the set of binary partition codes in order to produce a final binary image. This sixth step is known as integration.

The seventh step consists of testing the final binary image in one or more target architectures to verify that the system operates correctly and in the desired manner. In our case, the final binary image was tested on an emulator, and then on an embedded target used by the DT18 drone. The aim was to use the final binary image to carry out flight tests on drones. Details of the host hardware will be provided in Chapters 4 and 5. Note that the emulator was developed to facilitate migration of code developed on a local machine toward the software and hardware architectures used by Delair-Tech drones.

In summary, the seven steps are:

1) Partitioning (manual).

2) Modeling.

3) Transformation (from models to core source code).

4) Glue (from partitions to additional source code).

5) Compilation (from core source code and additional source code to binary code for the partitions).

6) Integration (from binary partition code to a binary image).

7) Execution (of the binary image on an emulator or real host).

2.2. Formal verification of models

The DO-178B standard recommends the use of test sets for validation purposes. The aim is to demonstrate the program's capacity to respond to input data in normal conditions. These test sets include the following operations:

– for time-linked functions, such as filters, integrators and lag, tests are carried out using multiple iterations in order to verify the characteristics of the function in context;

– for state transitions, test sets are carried out in order to implement the transitions which may occur in the course of normal operations;

– for program specifications expressed through logical equations, test sets need to verify the use of variables and Boolean operators.

The standard also prescribes robustness test sets, intended to demonstrate the program's capacity to respond to abnormal input data and conditions. It is important to note that the abnormal conditions in question are not events which are expected to happen, but unpredicted occurrences during the execution of the program.

Furthermore, these two types of test sets (classic and robustness) may not be sufficient, as they do not provide a 100 % guarantee that the program is error-free [BER 00]. It is very difficult, if not impossible, to carry out completely exhaustive testing using unitary tests. For this reason, the use of formal methods is recommended, as described below. This approach consists of using a verification method based on a formal model of the system to validate (e.g. using mathematical equations). While test sets verify that the system operates correctly in specific situations, the formal method verifies the adequation of the system with regard to the specification for any known or unknown situation. The application of formal methods is particularly important in validating the properties of aeronautical programs and UAS. Satisfactory results obtained using model checking techniques [BÉR 13] or statistical analysis through abstract interpretation [BLA 03] demonstrate the interest of using formal methods in a development cycle. In our case, we used automatic formal verification tools included in the MathWorks Simulink toolkit. These tools include Simulink Design Verifier, which uses formal methods to identify design errors in the models used by the programmer. It verifies blocks in the model which generate errors such as integer overflow, dead logic, array access violations and division by zero. It then generates a test case simulation, allowing the violation to be corrected.

2.2.1. *Model analysis*

Verification of model design is of crucial importance, notably to ensure the design quality of the different Simulink blocks and Stateflow diagrams used, which may contain potential errors (such as division by zero). A good modeling method should provide a modeling tool with criteria and guidelines for developing high-quality models. Guidelines may be expressed in the form of validated design rules or sample models.

Several authors have considered the development of statistical analysis techniques for verifying model behavior. Examples of contributions relating to this type of approach are described in [LIL 99]. The most widely used approach is model checking, which consists of listing all states through which the program may pass, then checking this list to validate property verification. Systematic model checking tools may also be used. These tools execute the program using input variables in order to satisfy test criteria. In our case, we used a model checking approach, with formal verification tools found in MATLAB/Simulink. These tools use formal proof approaches in order to test models, before handing them on to Embedded Coder for code generation, if the tests are successful, or generating an error report. These steps are shown in Figures 2.2 and 2.3.

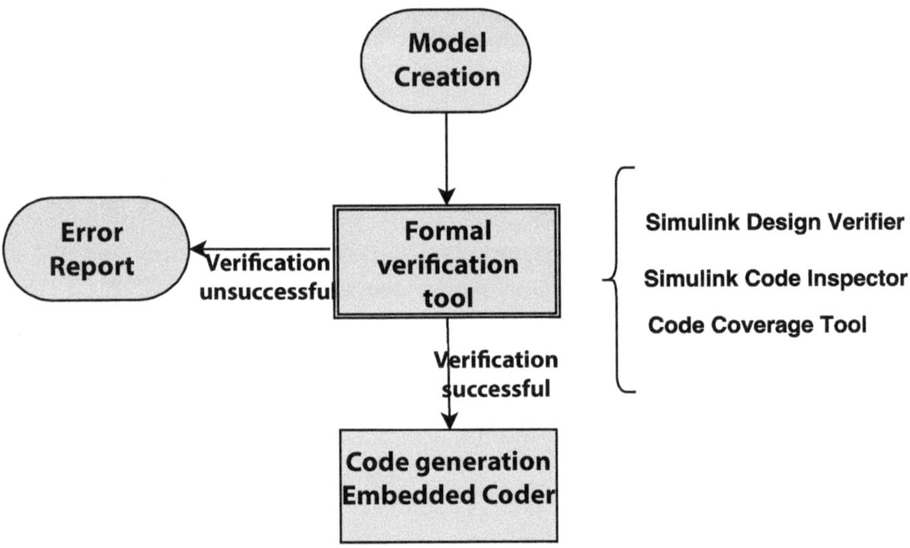

Figure 2.2. *Model verification process up to code generation*

In Figure 2.3, we see that our model successfully passed the verification operations carried out using formal verification tools. The report features four warnings relating to the range of variables used. These warnings give rise to manual corrections in the generated code. Figure 2.3 is a screenshot showing an example of model checking (carried out using Simulink Design Verifier) with the aim of obtaining suggestions for improvement.

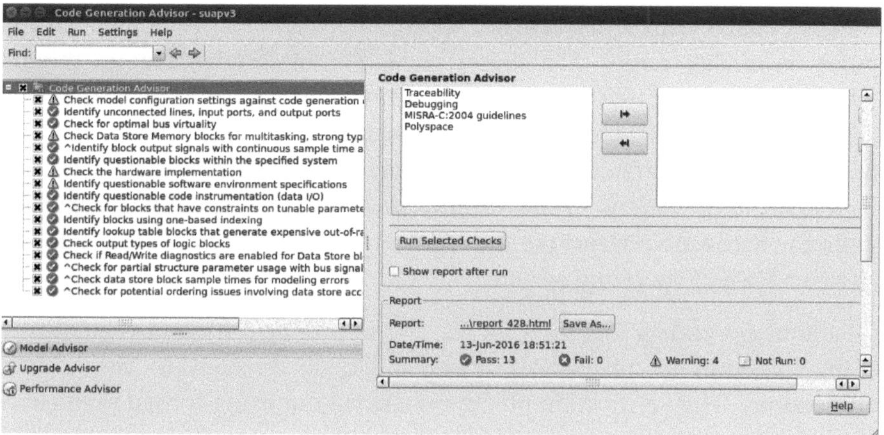

Figure 2.3. *Verification reports for our model. For a color version of this figure, see www.iste.co.uk/maxa/software.zip*

2.3. Advantages of MDD (Model-driven Development) methodologies

This methodology presents a number of advantages for software designers:

– the methodology speeds up the development phase, as models are used as the main artifacts in this step. The model is then transformed into high-level operational code. The use of models avoids the need to work directly on different lines of code. It is thus possible to obtain a rapid and operational representation of the code, speeding up the development process;

– applying the model concept allows greater abstraction of the program in comparison to traditional approaches. Each element of the model represents several lines of code. It is thus easier to gain an operational view of program architecture. Designers can use high-level models to verify and correct system behavior earlier in the process, minimizing the risk of incompatibility in the desired final product and reducing correction costs;

– decomposing embedded software into several partitions makes it easier to design. It isolates erroneous and insufficiently precise functions. This partitioning also simplifies the evaluation and validation processes, as safety assurances are easier to validate for modular systems than for single block systems [ALV 06]. The program is thus both modular and reusable;

– a key characteristic of the MDD approach is that the development process focuses on models rather than source code. The main advantage of this lies in the expression of the program as a model, using concepts which are independent of the host technology and closer to the specification. This makes models easy to specify, understand and maintain. It also makes them less sensitive to modifications in low-level technology: the model is independent of both the hardware platform (e.g. processor type or memory) and the logical infrastructure and operating system;

– automatic code generation increases the productivity and reliability of the final system. In most cases, generated code is ready to use with no modifications. The verification process is carried out using formal verification tools, guaranteeing the absence of undesirable properties. This verification is carried out iteratively during the model design phase in order to correct design errors as early as possible.

2.4. MDD contributions to UAS certification

In this section, we will analyze the requirements and state of the art of certification for a UAS system in order to demonstrate the relevance of model-driven methodologies. The need for certification relates to the integration of civil drones into national airspace. Certification constitutes a declaration of trust in the system. For software elements, the certification organization needs to be convinced that a rigorous development process has been followed. This objective requires the specification of rules and procedures for the physical and logical design of civil UAVs. These restrictions notably dictate the safety level which must be respected by drone manufacturers. In addition, the system must guarantee a level of reliability and fault tolerance. In response to these requirements, actors in the field of civil drone technology generally follow the approach taken in traditional civil aviation.

For an aircraft to be legally permitted to fly, it generally requires prior validation and certification by an authority such as the FAA (Federal Aviation Administration) or the EASA (European Aviation Safety Agency). The certificate of airworthiness covers a certain number of aspects relating to the design and operation of the aircraft, notably concerning:

– flight: performance, flight characteristics, stability, and control and responsiveness to handling;

– structure: payload, control surface, balance and stabilization surface and progression of wear;

– design: wings, control surfaces and the control system;

– energy source: the power supply;

– equipment: installation, electronics, safety equipment, etc.

For an aircraft to be certified, it must conform to its certificate type, i.e. the elements cited above must be fully controlled by the system in order to ensure safe operation. These rules are applied in civil aviation to regulate the airworthiness of commercial aircraft in the airspace. They may also be applied to the domain of civil UAVs, following the International Civil Aviation Organization (ICAO)'s Annex 2 [MAR 09], which states that operational standards must be respected whatever the nature of airworthiness of aircraft.

In France, the use of drones is governed by a decree issued by the Direction Générale de l'Aviation Civile (DGAC) on April 11th 2012, inspired by the regulations used for aircraft modeling. The regulations are divided into four different scenarios relating to different types of aircraft and for specified ground environments, at a given altitude, and following visual or instrumental flight rules.

These four scenarios are shown in Table 2.1. The regulation involves a risk assessment on the basis of several criteria: the mass of the aircraft, possible distance from the pilot, whether or not the aircraft is flying over an inhabited area, and the danger the flight poses to other users of the airspace. This approach is governed by two decrees published in April 2012:

– the decree of April 11th 2012 relating to the design of unmanned civil aircraft, the conditions of use and the capacities required for remote control;

– the decree of April 11th 2012 relating to the use of airspace.

In response to these criteria, drones must obtain certification for each usage scenario. Our aim is thus to contribute to eliminating the risks which may be caused by our secure routing protocol in the new UAS system within its usage scenario.

Scenario	Characteristics	Mission type
Visual flight regulations (VFR) scenario (S1)	– Drone less than 25 kg – Maximum distance of travel 200 m from controller – Maximum altitude 150 m	– Inspection of works – limited scale – Flight outside of inhabited areas – Visual flight
Instrument flight regulations (IFR) scenario (S2)	– Maximum altitude 50m – Drone less than 25kg – Travel distance up to 1km	– Flight outside of inhabited areas – Instrument flight – Inspection of works of surfacic or linear dimension
Visual flight regulations (VFR) scenario (S3)	– Drone less than 4kg – Drone equipped with equipment to protect third parties on ground (parachute, airbag, etc.) – Maximum distance of travel 100 m from pilot – Maximum altitude 150m	– Flight over inhabited areas (with authorization) – Inspection of a zone of known dimensions – Visual flight
Visual flight regulations (VFR) scenario (S4)	– Maximum altitude 150 m – Drone less than 2 kg – Unlimited travel distance	– Inspection of works of surfacic or linear dimension – Flight outside of inhabited areas – Flight with no visibility

Table 2.1. *Different usage scenarios featured in French regulations*

Certification authorities are currently carrying out research into developing a certification process for embedded software for civil drones. At the time of writing, the relevant authorities have yet to define a standard. However, a number of standards from the field of aeronautics (DO-254 [BAG 07], DO-178B, ARP-4761[1] and ARP-4754 [ARP 96]) have been applied to this area. As we have seen, most of the requirements which apply to aeronautics are also applicable to the case of drones. For this reason, we have considered the standard closest to our domain for the design of embedded aeronautics programs. Our intention is to follow the recommendations set out in this standard (the DO-178 family, particularly the most recent version, DO-178C

1 http://standards.sae.org/arp4761/.

[MOY 13]) to contribute to the certification of our routing algorithm, as and when a relevant standard emerges.

This standard governs the development and testing of embedded software used in commercial aircraft. It was written by representatives of aircraft manufacturers, software editors and certification organizations. It notably prescribes the use of a model-driven approach alongside certain code analysis methods. Applying this approach offers the following advantages:

– firstly, the guarantee that a verification procedure will be applied at each step in the development process, and the assurance of full traceability of requirements, from specification to implementation in the source code;

– next, the use of an MDD approach creates modular and reusable software. Modularity is particularly useful for certification authorities in the field of aeronautics: safety and security guarantees are easier to establish for modular systems than for monolithic systems [ALV 06];

– formal methods may be used to demonstrate the absence of design errors in the final system. Note that formal methods in certification standards are also considered in DO-331 [POT 12]. DO-331 was created in 2012 as an appendix to DO-178C and relates to the use of formal methods for the certification of aeronautical software. DO-178C focuses on the concept of model-driven development, while DO-331 introduces simulation and model coverage analysis.

Simulating a model allows us to validate an objective by analyzing the dynamic behavior of the system. The value and behavior outputs from the simulation should correspond to expectations. Model coverage is carried out during the simulation phase in order to compare specifications and models. Simulink Design Verifier and Simulink Code Inspector may be used for this purpose. The specificities of standards DO-178C and DO-331 are shown in Table 2.2, which also features examples of tools which respond to the requirements set out in the standards;

– DO-178C includes an additional document on the subject of tool qualification. This notion stipulates that certification authorities may validate tools, then accept their use by software designers. For example, they may be confident that a given automatic code generator will not create errors with regard to a model. In addition to speeding up the design process and providing automatic testing functions, these tools may also make the certification process easier. Several of the MathWorks tools which we used in the context of our project fall into this category, including Polyspace®, a statistical analysis and coding rule verification tool, which corresponds to the objectives laid out in the standard. Simulink Code Inspector also responds to five objectives cited in DO-178C: source code conformity to low-level requirements; source code conformity to software architecture; source code readability; traceability between the source code and requirements; and precision and coherency in the source code.

	DO-178C	DO-331
Content	– Modeling – Source code generation – Acquisition of executable object code – Unit testing of software partitions – Full requirement traceability	– Formal method – Model checking – Model simulation analysis – Model coverage analysis
Examples of tools	– MathWorks Model-based design – Polyspace – Simulink Code Inspector	– Model coverage – Simulink Design Verifier

Table 2.2. *Contribution of standards DO-178C and DO-331*

In conclusion, we see that to provide safety assurances for our secure communication architecture, we must follow all of the recommendations set out in standard DO-178C during the design process. In doing so, we contribute to certification of the UAS. Other advantages of this approach to developing critical embedded software include increased productivity, reduced costs and the added quality brought by the use of model-driven development techniques.

2.5. Choice of tools for applying MDD methodology

The content of the seven steps of the methodology was discussed earlier. In this section, we will identify the tools used for each stage in the process.

Step 1: organization of the secure communication architecture

As we have seen, the first step in the methodology consists of partitioning the functions of our secure routing protocol. This step is carried out manually, grouping specifications into blocks designed to represent partition classes. Our class diagram is shown in Figure 2.4, and will be discussed in more detail in the following chapter, dedicated to our proposed secure routing protocol.

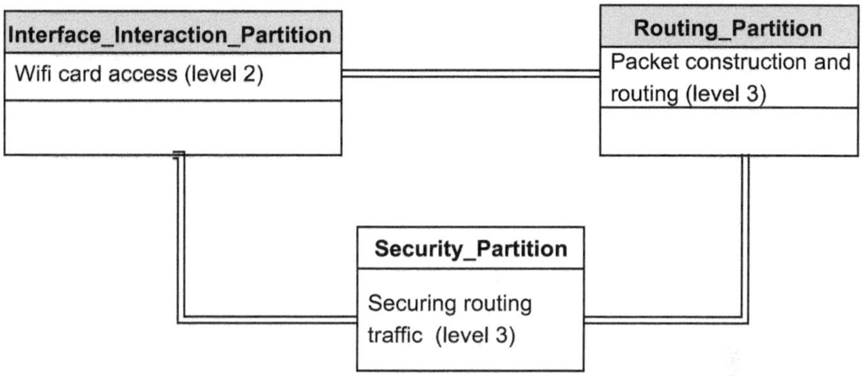

Figure 2.4. *Partition classes for our secure routing protocol*

The routing partition class is responsible for routing packets. The security partition class is intended to secure routing traffic, guaranteeing authentication. The interfacing partition class is given over to managing interactions between the classes which manage routing data at hardware input/output points. In practice, this liaison role consists of receiving incoming streams, depacketizing the data in the stream and transmitting it to the security and routing partition class. Each partition class is made up of function blocks. The routing block, for example, is made up of several sub-partitions, as shown in Figure 2.5. In this class of routing partition, the "Demux" sub-partition is responsible for retrieving packets which enter the system. The system then filters this input according to a clearly specified metric (e.g. the destination IP address or the sequence number) in order to transfer the packet to the corresponding block. We will not go into detail concerning the contents of this routing block here; it will be discussed in greater detail in the following chapter as it relates to the route discovery and maintenance mechanisms of our routing protocol.

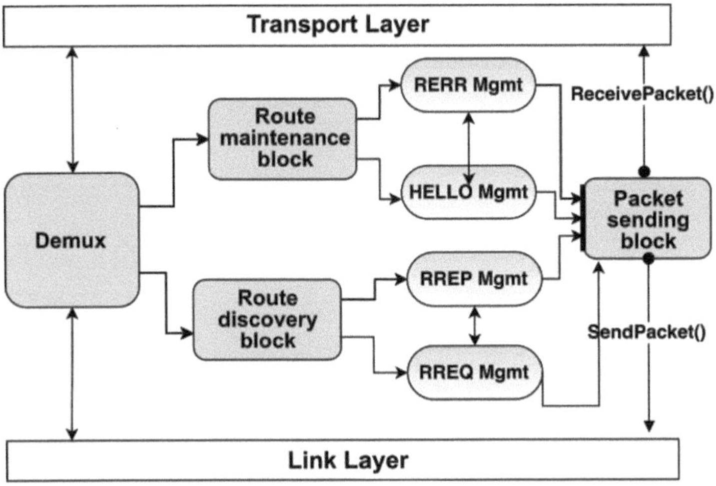

Figure 2.5. *Design of the routing partition class*

Step 2: modeling using Simulink and Stateflow

Design tools such as Lustre/SCADE or MATLAB/Simulink help us to manage the increasing complexity of embedded systems during the design process. We chose to use Simulink and Stateflow as they are recommended by industrial actors for use in association with standard DO-178C. Generally, once requirement partitioning has been carried out, designers are able to move on to the modeling stage. Using Simulink, we created a graphic representation of our system in the form of a block diagram. The blocks are interconnected by signals. Simulink blocks are also constructed on the basis of arithmetic operations or random signal generators, via operations written in MATLAB, C/C++, Fortran or Ada. In addition, Simulink allows users to simulate program operations and to obtain formal proofs from models.

Step 3: transformation of the model

The third step consists of translating models into a language which is as close as possible to that of the hardware. In our case, we used the MATLAB/Simulink Embedded Coder tool to generate code in C. Embedded Coder generates C and C++ code which can be used by embedded processors. This code generator creates high-quality optimizations, enabling fine-grain control of the generated code's functions, files and data. It also provides traceability reports, code interface documentation and automated software verification, corresponding to a software development process based on standard DO-178B. Before any code is generated, Embedded Coder uses a model checking tool, either to highlight design errors or to optimize the configuration of the model. The generated source code may also be optimized for the specific type of host processor used in the embedded system.

For our secure routing protocol, the Simulink and Stateflow models of partition classes were converted into 64 files, including one main file (5891 lines of source code in C). Figure 2.6 shows the output generated by Embedded Coder.

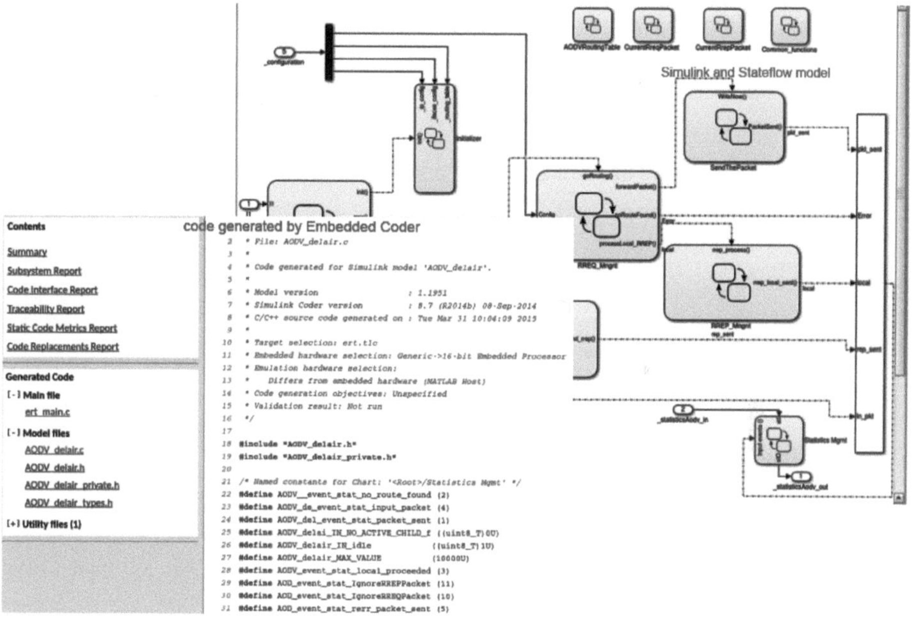

Figure 2.6. *Use of Embedded Coder to transform models into source code. For a color version of this figure, see www.iste.co.uk/maxa/software.zip*

Step 4: manual glue code

The code obtained in the previous step needs to be connected to the input and output of the system. The first procedure in this case involves managing the code for communications between partitions, which takes the form of signals in MATLAB/Simulink. The idea is to list these connections and to add the code corresponding to the input and output of each partition.

We also provided the embedded operating system with an input point for each partition. Interactions with the low-level interface are mostly carried out by the routing partition class.

Furthermore, glue codes specify different system configurations, as shown in Figure 2.7. First, the configuration of netlink sockets is used to establish communications between the user interface and the kernel level. This parameter is necessary in order to apply the appropriate security mechanism for each packet type. There are also kernel update codes for the operating system in conjunction with the state of the secure routing daemon at the application level. IP packetization and netfilter filtration are two basic methods of redirecting input packets to the corresponding output block. These different configurations are analyzed in greater detail in the following chapter.

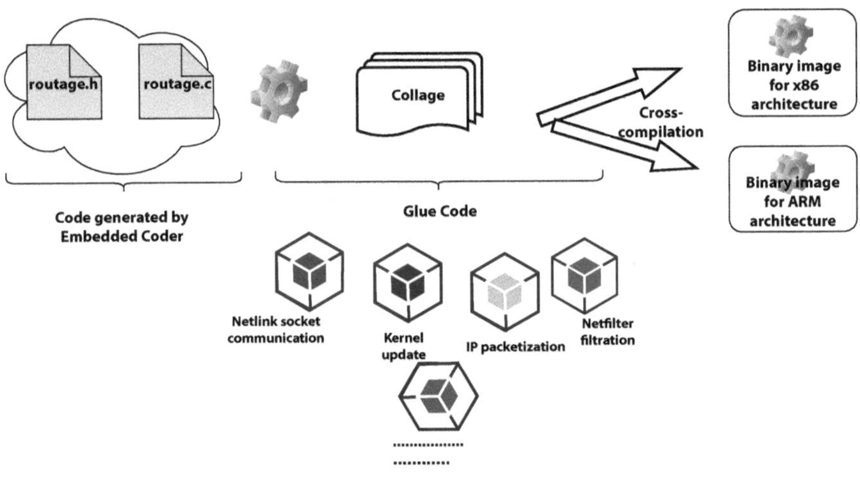

Figure 2.7. *Glue code connecting the operating system to the model. For a color version of this figure, see www.iste.co.uk/maxa/software.zip*

Step 5: compilation of C code using GCC and ARM GCC

The code generated by Embedded Coder and the glue code are independent of the processor used in the hardware platform. The final step is thus to obtain binary executables which can be used for each host hardware element. To do this, we used both the GCC compiler (version 4.6.3 for Ubuntu/Linaro) and an embedded compiler in the host hardware. Compilers are often used in association with a linker. They are responsible for the production of a binary image corresponding to the source code, using the configurations of the processor and hardware platform, and distributing the binary code to the appropriate addresses in the memory.

A further cross-compilation process needs to be carried out to enable the binary code to operate on our hardware, the PHYTEC board[2] used by the drones. The principle of cross-compilation is to compile programs on a machine with a different architecture to the destination hardware. The main reason for this is that it promotes local development of the software, with the addition of local modifications to obtain an executable binary. An additional reason is that, in our case, the local development machine was more powerful than the target; the destination system has limited processing power and memory resources. To this end, we used a toolchain (a set of instructions allowing the compiler to understand the target architecture) based on OpenEmbedded[3], as the source files for our embedded operating system are generated by OpenEmbedded. OpenEmbedded is a free framework for compiling software components intended for use in embedded systems, and was used to produce the Linux image used in our host hardware.

At the end of this step, we have a binary file ready for execution in a virtual machine using ARM architecture.

Tool chains

Table 2.3 provides a summary of the different tools used to apply the MDD method in this study. These tools offer certain advantages in terms of evaluation and certification. Functional safety evaluation is made easier by the

2 http://www.phytec.fr/produit/single-board-computer/phyboard-mira/.

3 http://www.openembedded.org/wiki/Main_Page.

modular approach taken during the partitioning and modeling steps. The use of qualified tools reduces or even eliminates tasks required during the certification process [VAR 13].

Step	Tools used
Modeling tools	Simulink & Stateflow
Transformation tools	Embedded Coder
Source code language	C
Compiling and link creation	GCC, embedded compiler
Embedded operating system	Linux generated by OpenEmbedded
Emulation/hardware platform	VirtualBox, ARM architecture

Table 2.3. *Tools used in our methodology*

Furthermore, these tools also contribute to software quality assurance. They ensure that the development of programs to be used in the UAS system, with all the processes involved, conforms to the plans and rules defined during the system specification and design phase. They also guarantee the respect of transition criteria for each process in the program lifecycle. Furthermore, they ensure conformity between specification requirements and the final software product.

2.6. AVISPA: a formal verification tool for security protocols

The specification of a protocol (and of algorithms in general) is often generated manually using mathematical formalisms. Simply believing a protocol to be fully reliable and being convinced of its ability to supply the expected security services is not enough without formal proof of these properties. We therefore need to carry out property verification before a protocol can be deployed and evaluated in a real environment.

In this section, we will consider the way in which the AVISPA tool may be used to create a specification and formal verification to highlight potential vulnerabilities in a protocol. AVISPA includes a graphics tool with the capacity to produce animations of protocol specifications, allowing users to vary execution scenarios in order to identify potential security failures.

The verification method used in this case is based on the use of formal methods.

Two methods are generally used to verify security properties: the "brute-force" method and the formal method.

The brute-force method considers cryptographic functions as operations over a set of bits. Security properties are described in terms of the probability that an attack will succeed. The intruder is represented by the Turing machine principle, with access to a set of information which may allow it to find a key and decrypt a message. Proof automation is difficult when using this approach, making it difficult to use in the context of our contribution to the SUAP protocol.

Using the formal method, however, the cryptography in question is considered to be perfect, following the Dolev-Yao model. This model allows us to affirm that an encrypted message may only be decrypted by applying the appropriate key. The intruder is modeled by a node which examines the network but does not have any cryptanalysis capabilities. In the formal method, the protocol description, security properties and the knowledge held by the intruder are explicitly defined by expressions. Formal verification tools thus test the security protocol in terms of its reactions to an intruder. If security failures are found in the protocol, the tools generate an example of an attack to illustrate the way in which the protocol does not conform to the desired security properties. This is the principle used in the AVISPA tool.

2.7. The need for verification

Successful communication between drones in a fleet is dependent on the formalization of the routing and security methods expressed in the SUAP protocol. This protocol describes the algorithm used by network nodes in sending and processing messages, along with the message exchange format. Several different verification tools have been designed for this purpose. The aim is to enable automatic proof of properties and to generate attack traces to highlight security failures. The different advantages associated with the use of formal verification tools are listed below, and the procedure in question is shown in Figure 2.8:

1) Identify known weaknesses: formal verification tools can be used to avoid the recurrence of known weaknesses. The utility of identifying these issues is questionable, since we are already aware of their existence. Nevertheless, it may serve to increase confidence in the verification tool

(AVISPA, in this case); furthermore, protocols containing these weaknesses are still being produced, despite widespread knowledge of their failures. The ping of death attack [HAR 99] on the ICMP protocol still works on certain IP telephones. This type of repeated error can only be avoided through systematic use of verification tools.

2) Identify attack variations: attacks which are identified automatically may not appear in their most usual forms. This may lead to an evaluation of the consequences of an attack.

3) Identify new weaknesses: the automatic verification tool is intended to highlight attack possibilities which correspond to unknown weaknesses. In practice, these weaknesses are rarely completely new, often coming from combinations of older attacks to which the system is not completely resistant.

4) Verify the utility of countermeasures: in our case, as we will see in the following chapter, the SUAP protocol includes several countermeasures (such as digital signatures and one-way hash functions) to secure communications. The hybrid Geographical Leaches approach is also used to counter wormhole attacks. These security mechanisms add to the complexity of the SUAP protocol, and may themselves introduce new vulnerabilities. It is therefore helpful to verify their utility.

5) Verify unpredicted cases: designers often work with assumptions concerning the execution environment of the protocol and the capabilities of the attacker. However, protocols may be used in other contexts, opening them up to a wider range of attacks. Automatic formal verification allows the identification of highly unlikely, but theoretically possible, attacks. In this way, weaknesses may be identified and noted in the initial assumptions in relation to conditions of use for the protocol.

2.7.1. *Why use AVISPA?*

A comparison of existing formal verification tools can be found in [PAT 10]. We decided to use AVISPA for the following reasons:

– Input language: AVISPA includes its own specific HPSL (High-Level Protocol Specification Language). This language is simple and close to the Alice and Bob formalism which we use to describe protocols.

– Comprehensible output: AVISPA presents results in graphic form and generates comprehensible attack traces. This makes it easy to verify the properties of the SUAP protocol.

– Properties: AVISPA covers the properties which we wish to verify in relation to message authentication.

– AVISPA is also compatible with the SPAN graphics tool which generates graphic representations of attack traces. SPAN (Security Animator for AVISPA) is designed to support the design of formal specifications. It allows the creation of animations of HLPSL specifications, i.e. interactive production of MSCs (Message Sequence Charts) which may be seen as “Alice and Bob” traces of HLPSL specifications.

– Expressivity: AVISPA is not an expressive tool, making it easy and ergonomic to use. Expressivity relates to the representation of different types of execution variables during verification, such as the communication channel or indeterminism (i.e. the notion of choice or chance). Our aims in this case did not include the creation of a verification tool or the development of add-ons to improve existing tools, so we selected our tools based on ease of verification and speed of use.

– Availability: AVISPA is free and available to the general public.

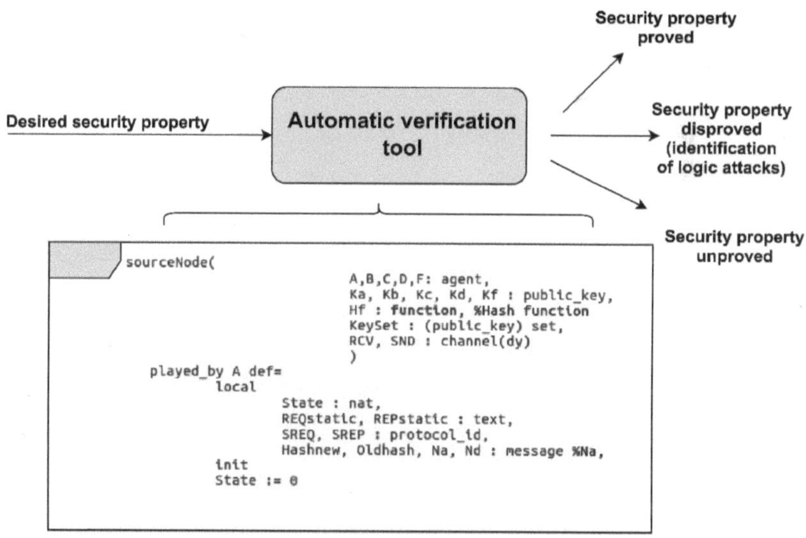

Figure 2.8. *Automatic verification procedure. For a color version of this figure, see www.iste.co.uk/maxa/software.zip*

2.8. Additional tools: simulation and experimentation

The context of UAS is complex and must be considered in parallel with the development of functionalities. It notably generates particular requirements during the design process, independent of the functions in question. These requirements are said to be non-functional; they concern safety and security properties linked to the selected properties. They are treated using formal design verification tools in conjunction with Simulink and Stateflow. Security properties are verified using AVISPA, as described above.

The three following software partitions are affected by functional requirements, which must be validated through both simulation and real-world experimentation:

– the routing partition class, responsible for the implementation of routes between different nodes in a network;

– the security partition class, which manages routing packet security (authentication and integrity);

– a security extension class, associated with protection against wormhole attacks.

2.8.1. *Testing and validation using emulation and network simulations*

Our aim here is to evaluate the performance of different routing protocols in the context of a proposed scenario. In the case of our study, this scenario involves three drones on a surveillance mission. If the three drones circulate according to their flight plan, an obstacle or limited range may be able to prevent a drone from communicating directly with the base station. In this case, one or more of the drones need to be able to act as relay points for communications with the ground unit.

This basic scenario is shown in Figure 2.9.

2.8.1.1. *Reasons for using emulation and simulation*

Several solutions are available to evaluate the performance of an ad hoc routing protocol. In existing literature, these fall into two types:

– evaluation of protocols using a simulation tool: this solution enables the use of multiple parameters associated with the environment and with mobility, and also allows us to evaluate the implementation of a protocol produced by the simulator. Implementations for a specific operating system cannot be tested using this solution;

– testbeds: this is an ideal solution, allowing us to evaluate an implementation in a real-world situation, but is very costly in terms of both time and materials.

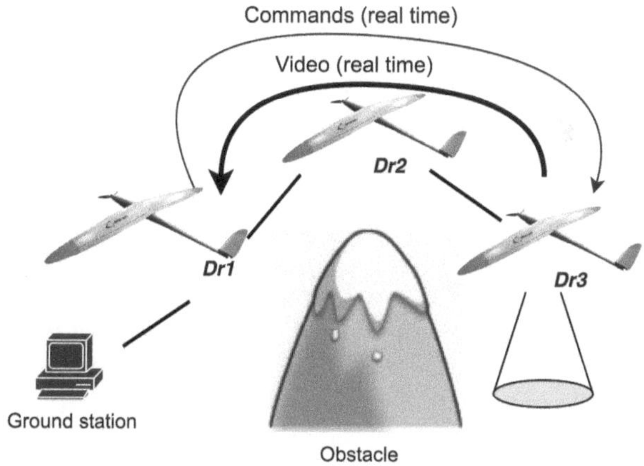

Figure 2.9. *Basic scenario used for our experimental protocol. For a color version of this figure, see www.iste.co.uk/maxa/software.zip*

A comparison of these solutions is shown in Table 2.4.

Simulation	Testbed
+ Fast	– Time-consuming
+ Scalable	– Limited scalability
+ Mobility	– Limited mobility
+ Reproducible	– Not reproducible
+ Controlled environment	– Undesirable interference
– Different code from system software	+ Real implementations
– Approximation of the real environment	+ Real environment

Table 2.4. *Advantages and disadvantages of simulations and testbed evaluations*

Both paradigms therefore merit consideration. Our aim is to evaluate an implementation for a standard operating system without having to carry out real-world unit tests. Another solution for the evaluation of ad hoc routing protocols involves the use of virtual machines. While variants are available for a range of parameters, in this case, we will focus on a single scenario.

2.8.1.2. *Experimental protocol for UAANET networks*

Our aim is to identify the routing protocol which offers the best performance for a drone fleet in a realistic context of use. Our choice must be made in such a way as to maximize the quality of service for users, e.g. in terms of real-time video for surveillance missions, considering elements such as display latency, video quality and service availability.

In studying the state of the art of existing MANET protocols, we noted a difference between the assumptions considered in simulation studies in a MANET environment and the real conditions of use of an UAANET network. For example, many of the tests carried out on MANET routing protocols involve large numbers of nodes, generally between 50 and 200; this does not correspond to the situation here. Furthermore, few studies have been carried out specifically for drone networks, which present very specific characteristics in terms of mobility. Finally, many contributions are based on results obtained through simulation, making abstraction of a range of phenomena which may arise in testbed situations.

2.8.1.3. *Architecture of the testing tool*

An overall representation of our hybrid tool for testing routing protocols is shown in Figure 2.10.

The details of this tool may be found in [MAX 15]. In this section, we will simply outline the main modules.

– Hypervisor: the choice of hypervisor was essentially based on the available network possibilities. We chose to implement our routing solution in real time using a Linux system. We opted for *VirtualBox* [WAT 08], an easy-to-use tool which allows communications to be established between multiple virtual machines over a virtual network.

– Network virtualization: for a network to be established, each of the virtual machines requires a unique interface in a shared network, as in the case of wireless networks. *VirtualBox* allows the creation of these connections via an internal network system, designed to interconnect multiple virtual machines. To simulate the loss of connections which may occur in realistic implementations of UAANET networks, we included a mobility element in the experimental protocol via the virtualmesh framework [STA 11]. This framework allows interfacing between a Linux system and an OMNeT++ simulation[4]. Its integration into the experimental protocol is shown in Figure 2.11. OMNeT++ is a network simulator coded in C++. Being modular, it can create simple simulations of a wide range of standardized systems and protocols using the INET framework [STE 11].

– Mobility: we used traces provided by the Delair-Tech company to recreate a mobility scenario corresponding to scanning a zone, in the aim of creating the most realistic scenario possible. Using a real trace, rotations and scale changes, we were able to recreate a scenario with three drones working together to scan a zone.

– Access to the medium: this function is provided by the Virtualmesh framework, used alongside OMNET++ and the INET module. Virtualmesh thus allows us to simulate a realistic medium access situation (with collisions, signal damping, etc.).

– Implementation of selected protocols: the protocols selected for the first evaluation series were OLSR, AODV and DSR. These pre-existing and widely used protocols have been discussed in a number of publications and have formed the basis for many further developments [SAH 14]. Another motivating factor in the choice of protocols was the availability of implementations for Linux.

Moreover, note the existence of a third major family of routing protocols which could have been tested; these are the geographic routing protocols, of which GPSR is a notable example. This family of protocols uses geographic positions to relay packets. Localization information is used to determine the next hop on the path to a destination.

4 OMNeT is a network simulator with the capacity to simulate mobile ad hoc networks: https://omnetpp.org/.

– Profile of generated application traffic: we extracted the main characteristics from real traces, corresponding to traffic emitted during a mission with a direct link. Using extrapolation, we were able to generate similar traffic for our three nodes, the characteristics of which are shown in Table 2.5.

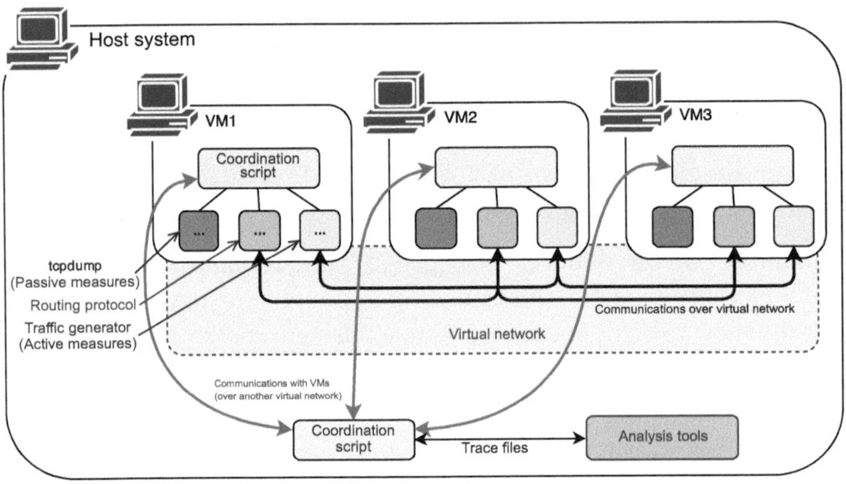

Figure 2.10. *Global view of a system designed to test routing protocols by emulation. For a color version of this figure, see www.iste.co.uk/maxa/software.zip*

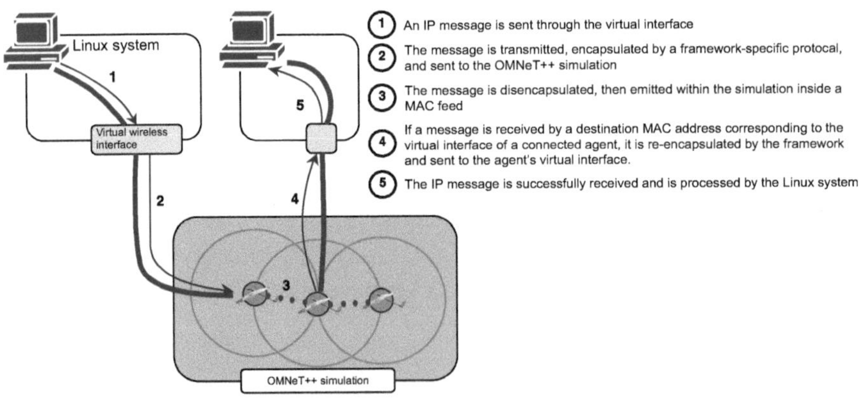

Figure 2.11. *Sending an IP packet through the* Virtualmesh *framework. For a color version of this figure, see www.iste.co.uk/maxa/software.zip*

Type	Source→Destination	Packet size	Frequency
Tick	Dr1→Dr2, Dr1→Dr3	64 bytes	1.0 packet/s
Georef	Dr2→Dr1, Dr3→Dr1	64 bytes	1.8 packet/s
Command	Dr1→Dr2, Dr1→Dr3	64 bytes	0.034 packet/s
Video	Dr2→Dr1, Dr3→Dr1	4 Mbit/s	

Table 2.5. *Generated traffic*

The traffic under consideration in terms of useful data corresponds to HD video which might be generated in the case of surveillance applications. We selected the H264 codec, widely used for video of this quality. As H264 is a variable bitrate codec, we selected an average rate of 4 Mbit for a full HD image (1920x1080 pixels) at 30 images/sec. These characteristics were extracted from a promotional video produced by Delair-Tech, essentially made up of in-flight images captured by a drone. We introduced a uniform variability, arbitrarily set at 50%, for image size. The equivalent of one image was sent split into 1000-bit packets in order to avoid fragmentation.

2.8.2. *Testing and validation using real experiments*

To evaluate the behavior of a routing protocol in a real environment, we must be able to consider its execution in the final embedded system. The aim of these experiments is to check for conformity between the specification, the high-level model and the final execution of the generated code.

For our study, we validated the routing and security functions of our routing protocol using real-world tests. These experiments were carried out using Delair-Tech DT 18 drones [DRO 16].

Our experiment required the use of appropriate evaluation topologies for the different systems involved in order to validate each partition of the routing protocol. The common topology in question, presented in the following chapter, is shown in Figure 2.12.

The drone system used for our test platform included the following elements:

– three DT18 drones;

– three pilot stations associated with each drone. Note that only one of these stations (station 1 in Figure 2.12) received payload data from the nearest drone.

The other stations (e.g. the "station 2" and "station 3" nodes in Figure 2.12) are only included in the network to respond to French regulations for civil drones, requiring a security connection to each drone at any given moment (to protect against malfunctions in the UAS);

– three supervision programs for mission control (flight plan tracking and interactions with environmental components).

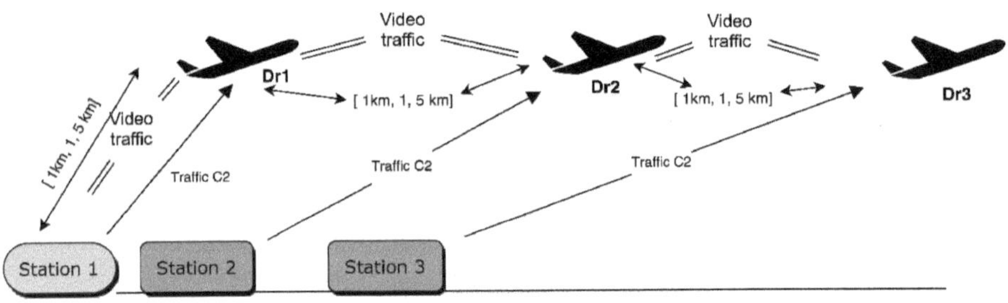

Figure 2.12. *Common topology for real-world tests. For a color version of this figure, see www.iste.co.uk/maxa/software.zip*

The interactions between these sub-systems are shown in Figure 2.12. In this figure, we see that a single liaison is used for communications between each drone's autopilot and a control station. In this topology, payload information is sent from drone Dr3 to station 1 via the neighboring nodes (Dr2 and Dr1). Note that only the information acquired by the payload is relayed across the network in the context of our real-world tests. Command traffic is sent to each drone directly.

3

Application to Communications in a Drone Fleet

3.1. Introduction

A communication architecture defines all of the entities required for communication along with the rules governing exchanges between entities. These sets of rules form the basis for the creation of a network of interconnection and the implementation of network services. In the case of a drone fleet, a communication architecture must be defined and implemented in order to manage the exchange of information flows between the control station and drones. A drone fleet may be defined as a set of cooperating UAVs operating together to share tasks and carry out complex missions (e.g. by extending the range of communications by relaying data). Other categories of communication network may be considered for use within a drone fleet. In this chapter, we will highlight the advantage of an ad hoc drone network (UAANET). We will then present a study of the security problems encountered in UAANET networks. This contribution is justified by the presence of weaknesses inherent in the ad hoc architecture, which could be particularly dangerous in the context of a drone fleet (e.g. the risk of a fleet member being hijacked). This study notably allows us to identify the security mechanisms needed to secure the communication architecture. We shall also present the different mechanisms used in a new routing protocol that we have developed, known as SUAP (Secure UAV Routing Protocol). The model-driven development of the mechanisms in the protocol and the associated formal verifications will be covered in detail.

Figure 3.1 offers a summary of the different themes discussed in this chapter. The first element of research work presented consisted of choosing an initial protocol. This choice was based on an evaluation study carried out using simulation and emulation tools. Next, security blocks were added to the routing protocol in order to secure communications between members of the drone fleet. Like the routing protocol, these security mechanisms were designed using a model-driven approach and verified using formal verification tools, as described in the previous chapter. This constitutes an improvement to the program code, and thus contributes to the certification of the final UAS system. Lastly, the final system was evaluated using simulation (and emulation) tests and through real-world tests using multiple Delair Tech DT18 drones.

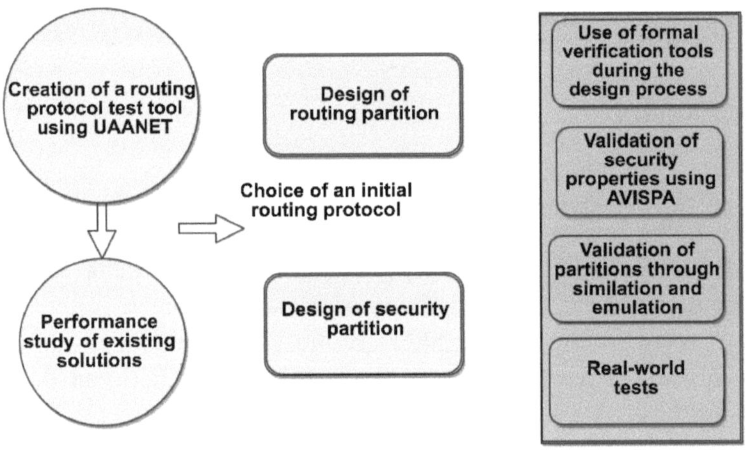

Figure 3.1. *Design and verification of the SUAP protocol. For a color version of this figure, see www.iste.co.uk/maxa/software.zip*

3.2. Cooperating unmanned aeronautical systems

A drone is an unmanned aircraft equipped with an embedded autopilot[1], which may be controlled remotely from a station on the ground. Two terms are generally used in literature to denote these systems: UAVs and Remotely Piloted Aircraft Systems (RPAS). Unlike UAVs, RPAS require the presence of

1 An autopilot is a hardware and software module offering piloting assistance.

an active remote pilot on the ground, but do not necessarily feature an onboard autopilot. The term UAV specifically refers to drones with active embedded autopilots.

Drones may carry out a variety of missions; in the past, these were mostly of a military nature (target surveillance, reconnaissance, etc.), but more recently, their use has extended into the civil domain (e.g. for audiovisual, industrial or agricultural purposes). In this book, we shall only consider the case of civil drones.

Civil drones contain multiple sensors and carry a payload.

These elements contribute to a drone's autonomy and allow it to be used for a variety of applications. In the audiovisual domain, for example, piloted airplanes or helicopters were traditionally used in creating reports. These missions could be dangerous and poorly suited to piloted aircraft and also represented a significant cost. Tasks of this kind can now be carried out by civil drones, adapted to different mission types and conditions.

Drones receive control and command (C2) traffic from the ground station at a pre-defined frequency. They also send configuration information relating to flight conditions (position, speed, etc.) and data acquired by the payload back to the ground station. These data allow the ground operator to monitor flight and to intervene where necessary by sending commands to the drone.

3.2.1. *Unmanned Aircraft/Aerial Systems*

Unmanned Aircraft/Aerial System (UAS) is a term used by the US Federal Aviation Administration (FAA) to cover the different denominations shown in Table 3.1, providing an umbrella term with the aim of encouraging use of a system principle. In this case, the UAS is made up of drones, different communication and data transfer connections, one or more ground systems, and additional mission security systems. These last systems are add-ons intended to increase mission reliability and to respond to certification requirements. Note that both the FAA and the European Aviation Safety Agency (EASA), operating through the Direction Générale de l'Aviation Civile (DGAC) in France, insist that UAS should be controlled from the ground at all times [DAL 08]; this restriction clearly applies to the use of drone fleets. In order to fulfill these requirements, a specific ground station is often required for each drone in the fleet [MAX 16c].

Name	Acronym
UAS	Unmanned Aircraft System
UAV System	Unmanned Aerial Vehicle System
RPAS	Remotely Piloted Aircraft System
RPV System	Remotely Piloted Vehicle System

Table 3.1. *Different terms used to denote UAS*

3.2.2. *Payload*

The payload of a drone is the set of embedded systems which allow it to carry out its assigned mission, for example, a camera, video camera, multi-spectrum camera, thermal imaging camera, probe or any other type of sensor (used to measure air parameters, for instance, such as temperature, pressure or pollution levels). Drone systems may thus be used to inspect a given zone, map an area, take thermal measurements for factory buildings, etc.

3.2.3. *Ground station*

The ground station is a set of hardware and software elements used to control the movement of drones. Depending on the type of station used, they may be equipped with user interfaces to allow ground operatives to monitor the position of a drone in real time, using a map to display the route taken by the drone. They may also permit users to configure the altitude of the drone and the parameters of the payload.

Using a real-time connection between the ground station and the drone, the supervision program (often known as the Ground Control Station or GCS program) allows operators to view video or photos taken by the drone. All data treated by this program are tagged with a geographic location and may be saved or shared with other geographical information systems.

3.2.4. *Drone fleets*

From the start of this chapter, our discussion has focused on the operations of a system of mini-drones and on different types of civil applications. We have seen that the use of these mini-drones creates added value in terms of performance and productivity. However, the capacity of a single-drone system is limited in terms of both space and time [BEK 13]. For missions over several hours and/or covering large areas, use of a single drone may be limiting due to battery life and range considerations (notably if obstacles are present). The use of multiple drones operating in collaboration presents a number of advantages:

– Cost: the cost of obtaining and maintaining mini-drones operating as a fleet may be less than that of a single large drone.

– Scalability: a single drone can only cover a limited area. The extent of a mission depends on battery life and the presence of obstacles (e.g. a mountain). Using a fleet, relay techniques may be used to get around obstacles and increase the zone covered during an operation. This function will be discussed in greater detail in this chapter, notably in relation to the concept of ad hoc drone networks, which are used to develop an autonomous communications network.

– Mission success: missions may need to be aborted if a physical or software element of a UAS fails. Using a fleet, however, means that a mission may still be able to continue if a drone fails or runs out of battery, by reconfiguring fleet resources.

– Latency: [YAN 10] shows that node density is proportional to mission execution time. The more nodes there are in a network, the less time will be required to carry out a mission.

– Network resources: drone fleets enable the reuse of network resources, notably bandwidth. When using a single drone, a certain quantity of bandwidth needs to be allocated for the whole duration of a mission.

3.3. Ad hoc communications architecture for a drone fleet

A wireless ad hoc network (as shown in Figure 3.2) is characterized by a set of potentially mobile entities with one or more radio interfaces that establish a temporary communications network in response to the requirements of a specific application. These nodes may enter or leave the network at any time. Wireless ad hoc networks are decentralized and are able to self-organize without requiring a fixed infrastructure. If an emitter is not within the direct range of a destination machine, information is transmitted through neighboring nodes along a chain established and maintained by the network in cases of changing topology.

The main advantages of ad hoc architectures in relation to other types of architecture for a drone fleet are:

– Scalability and agile reconfiguration of missions carried out by a drone fleet: using a system based around a central structure or on satellites, the zone of operations is limited by network coverage. Furthermore, communications between drones may be blocked by the presence of obstacles. Ad hoc architectures make it possible to establish a chain of drones and thus circumvent the obstacle.

– The reliability of communications in an unmanned aeronautic system: drones exchange messages relating to commands, configuration and mission operations in real time. These messages are exchanged in real time in a dynamic environment, in which communication links fluctuate and may be broken. Furthermore (depending on specific mission characteristics), drones may enter and leave the communications network. The self-organizational capacity of wireless ad hoc networks allows drones to look for new pathways if a link in the existing chain is broken.

– The use of ad hoc networks also means that smaller telecoms hardware elements can be used. In a fixed infrastructure network, each drone must carry an emitter, a receiver and heavy, bulky communication aerials. In an ad hoc drone network, each aircraft only carries the equipment needed to recognize its neighbors. The miniaturization of equipment also increases the endurance of the system by reducing energy consumption.

– No need for a dedicated infrastructure: wireless ad hoc networks differ from other network types in that they do not include a centralized infrastructure. Network nodes themselves are responsible for creating and maintaining network connectivity.

– The use of ad hoc wireless networks guarantees mobility and flexibility [CHE 13]. Dynamic topology means that the mobile nodes in the network can move freely and arbitrarily in relation to mission objectives. As topological links may be mono- or bi-directional, nodes are free to enter or leave the network.

– The security element of a mobile ad hoc network is distributed between the nodes, rather than being concentrated in a centralized and cellular network.

This means that there is no one vulnerable point in the network. Responsibility for network integrity is delegated to each node in the network, and each node is able to authenticate exchanged messages through a secure routing mechanism.

3.3.1. *Ad hoc drone network*

The ad hoc drone network (shown in Figure 3.2), known as a UAV Ad Hoc Network (UAANET) or a Flying Ad Hoc Network (FANET), is a subcategory of the MANET mobile network. It involves the deployment of a fleet of drones and ground stations through a wireless ad hoc network. Drones work together and with the ground control station(s) to exchange data, which may relate specifically to routing (control packets) or contain information specific to the UAS.

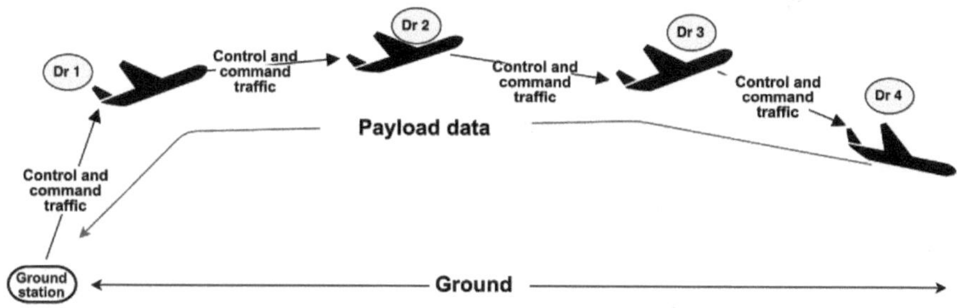

Figure 3.2. *Ad hoc drone network. For a color version of this figure, see www.iste.co.uk/maxa/software.zip*

Several types of messages may be exchanged between drones and the GCS. These different application flows are listed in Table 3.2.

Flow type	Connection type	Status
Configuration (autopilot launch, payload settings, etc.)	Connected upload	Critical
Ticks, joystick	Disconnected upload	Non-critical
Georeferencing	Disconnected download	Non-critical
Data (photos, video)	Disconnected download	Non-critical
C2	Connected upload	Critical

Table 3.2. *Different types of application flows in an ad hoc drone network*

These application flows generally include:

– Georeferencing of the drone: drones send this information to the ground control station periodically. It allows real-time determination of the drone's position. This flow does not require a specific quality of service, and some loss or latency is not critical.

– Tick flow (or heartbeat): this signal flow is sent periodically by the GCS in order to communicate with the drones. It is used to test the state of the connection between the ground station and the drone. It does not require a specific quality of service and intermittent packet loss is acceptable.

– Control and command (C2) flow and configuration: these are commands sent to the embedded autopilot or the payload. This type of flow is considered to be critical as it needs to be exchanged and processed in real time. It therefore requires minimum delay and maximum reliability in the communication links.

– A data flow: like configuration flows, this type of traffic has strict requirements in terms of loss rate and delay. It is traditionally sent from the drone which is furthest from the ground station and relayed by the members of the drone fleet.

UAANETs are generally part of the MANET family of networks. They do not require a fixed infrastructure and are based on nodes working together to exchange data. However, UAANET networks have certain specific characteristics which set them apart from other types of MANET network.

3.3.1.1. *Network connectivity*

Connectivity within UAANET networks is often intermittent due to the movement of drones, which results in temporary disconnections. These movements are governed by a ground operator or by pre-loaded flight plans. Thus, communications between a node and its neighbors may be interrupted by any position change. The communications connection from a source to a destination may become unavailable for an unknown period of time; the network therefore needs to react to this situation, finding a backup pathway and avoiding any significant loss for application flows.

3.3.1.2. *Node density*

Node density may be defined as the average number of nodes in a given geographical area. UAANET networks discussed in published literature generally only include a few drones [DAN 09, CAM 10]; they differ from MANET networks in this respect. The high speed at which drones can travel allows them to cover large areas, meaning that fewer aircraft are required.

3.3.1.3. *Energy consumption*

The life expectancy of an ad hoc drone network depends on battery autonomy and on the energy consumption of the captors involved. If one of the batteries of one of the nodes becomes flat, the node must abandon the mission, leading to a modification in network topology. Depending on the type of drones in question, this issue may arise in UAANET networks, notably if Paparazzi-type drones[2] are involved, as they only have 10–15 minutes of flight autonomy [PAP 03]. However, in most cases, and notably for Delair-Tech drones [MAN 14], which have around 2 hours of flight autonomy, this issue is less significant [MAN 15]. In this case, network topology may be maintained over a longer period.

3.3.1.4. *Strict time constraints*

UAANET networks are generally used for real-time applications (such as surveillance missions). Control and command flows must therefore be processed by the drones in real time in order to avoid losing control of the aircraft.

Table 3.3 summarizes the main specificities of the UAANET network.

	UAANET	**MANET**
Node mobility	very fast	slow (depending on nodes)
Mobility model	specific models, e.g. PPRZM	generally random
Node density	low	high (depending on application type)
Topology modification	fast	slow (compared with the UAANET network)
Radio propagation model	at altitude, generally in LoS (Line of Sight)	generally on the ground, no LoS (following network topology)
Flow type	real time	non-real time

Table 3.3. *Comparison of UAANET and MANET networks*

2 A complete system of open source software and materials used to manage unmanned aerial systems. It consists of an autopilot to direct the aircraft and a ground station containing mission planning programs.

3.4. Routing protocols in an ad hoc drone network

Routing is a method used to transmit data from an emitting node to a recipient or recipients. In mobile ad hoc networks, routing is based on a packet re-sending approach. The challenge here lies in identifying an optimal pathway. Routing involves calculating the best pathway between two given nodes in a network. A certain metric thus needs to be assigned to connections so that the route identification process may, for example, simply involve calculating the shortest pathway between a source and a destination.

To reach its destination, a message may be relayed by intermediary nodes, according to the availability of communications links. In the specific case of an ad hoc drone network, drones must relay workload traffic (and, in some cases, control traffic, according to the applicable regulations) between each other and the ground station(s) in the context of their missions. To this end, nodes use a routing table that guarantees correspondence between a destination node and a pathway to this node. Proactive or reactive techniques may be used. In the first case, nodes establish a route before a message needs to be sent at application level; in the second case, this route is established as the need arises. An example of a routing mechanism is shown in Figure 3.3.

Note that the high level of node mobility is a particularly important feature of ad hoc drone networks. Frequent changes in topology lead to route changes and create a management overhead in the routine mechanism. Lost routes need to be restored rapidly in order to avoid packet loss or sub-optimal bandwidth usage resulting from retransmission. As the traffic is critical, a rapid and effective backup mechanism is necessary. These constraints mean that a routing protocol needs to have excellent overhead properties and quasi-real-time execution. A number of routing protocols for ad hoc drone networks have been published in response to this context; most are extensions of routing protocols used for MANET networks [SAH 14].

Ad hoc routing protocols may be classified according to a variety of design metrics: route creation and maintenance methods, or data transmission methods. In [MAX 17], we presented a study of existing UAANET routing protocols, along with a discussion of the security challenges associated with the use of UAANET networks. In what follows, we shall simply present the main families of UAANET routing protocols without going into technical details. Interested readers may wish to consult [MAX 17] for further information in the context of routing protocol security issues.

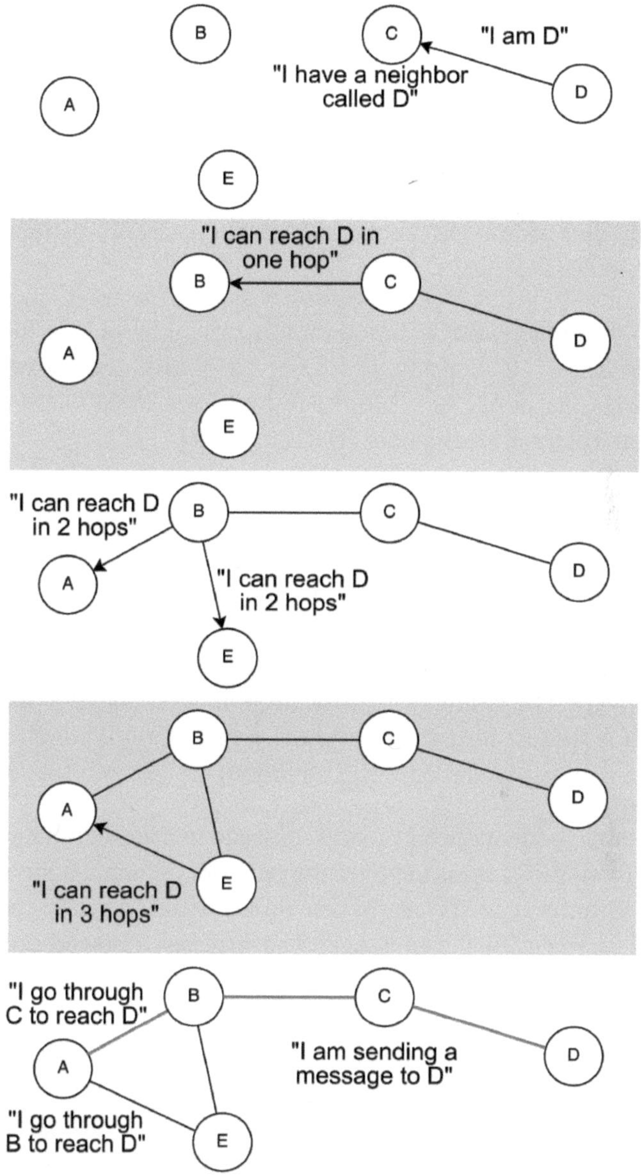

Figure 3.3. *Illustration of an ad hoc routing mechanism establishing a route from A to D. For a color version of this figure, see www.iste.co.uk/maxa/software.zip*

3.4.1. *Hierarchical protocols*

Hierarchical routing protocols operate by assigning specific roles to nodes, roles which vary over time. Certain nodes are selected and fulfill specific functions, leading to a hierarchical vision of network topology. For example, selected nodes may act as bridges for a certain number of other nodes that are attached to it. This simplifies the routing process, as only pathways between bridges need to be identified.

UAANET routing protocols based on this principle include the Clustering Algorithm of UAV Networking [KES 08], Mobility Prediction Clustering Algorithm (MPCA) [ZAN 11], Multi Meshed Tree (MMT) [MAR 11], and the Disruption Tolerant Mechanism (DTM)[FU 07].

3.4.2. *Reactive protocols*

The core principle of reactive routing protocols is to only remember pathways which are currently in use for the routing process. Pathways are only constructed in response to a data sending request (generated by the application layer). The pathway identification method consists of flooding the network with a request in order to find the target station, which responds by following the pathway in the opposite direction.

When a source node wishes to send a message to a destination node, it sends a route request (RREQ) message over the network. When a node receives this request, it transmits it to its neighbors, unless it has already received or is the intended recipient of the message. The process followed during request transmission varies according to the type of protocol in question. In routing table protocols, nodes update their routing tables by adding an entry indicating the next hop needed to reach the source node indicated in the RREQ, creating a reverse route back to the source. In the case of source routing protocols, such as DSR, intermediary nodes add their identifiers to the request messages they received, creating a record of the route taken from the source node. This route is then stored in the request transmitted to the recipient. In both cases, on receiving an RREQ, the destination node responds with a route reply (RREP) message. This packet travels back along the route taken by the RREQ message, either using the routing tables of intermediary nodes or by following the route indicated in the RREQ header. When using routing tables, on receiving an RREP message, the source node adds an entry to its table indicating the route taken to reach the recipient in question. The protocol must also ensure periodic verification of connectivity over each connection.

The drawback to this type of protocol is that they must be able to react quickly to identify an alternative pathway if a route is cut off during use. These maintenance mechanisms are costly in terms of management time and create peaks in control traffic every time a route needs to be identified or repaired. They result in lagging and in data packet buffering in cases of route failure.

Protocols based on reactive principles include the following broad families: Ad Hoc On Demand Distance Vectors (AODV) [PER 03a], Dynamic Source Routing (DSR) [JOH 01] and Temporally Ordered Routing Algorithms (TORA) [GUP 10], along with their variants: Time Slotted AODV [FOR 07], Rapid-reestablish Temporally Ordered Routing Algorithm (RTORA) [ZHA 13] and Reactive-Greedy-Reactive (RGR) [SHI 12].

3.4.3. *Proactive protocols*

Proactive routing protocols are based on the principle of memorizing routes to each node in a network. Network nodes are responsible for maintenance via the periodic exchange of monitoring messages. Periodic routing data updates are transmitted by the nodes in the network. This exchange allows nodes to collect all of the information they need to calculate routes according to pre-defined criteria. These criteria include the shortest route technique (smallest number of hops between nodes), transmission time or the bandwidth available for the identified route. A cost may thus be assigned to each communication link, which may reflect the rate of use, relative communication delays, or any other factor for minimization during the data routing process.

Proactive routing protocols use either a link state algorithm or a distance vector algorithm. In the "link state" method, each node retains a vision of the whole network topology by means of periodic requests checking the state of links to neighboring nodes. To ensure that this vision is up to date, each node periodically transmits the link states of its neighbors to all nodes in the network, even when a link state has changed. On receiving link state information, each node updates its vision of the network topology and applies an optimal route calculation algorithm in order to identify the next node for any given destination. Dijkstra's algorithm is one of the most widely used methods used in the shortest pathway calculations [SKI 90]. Note that the routing node calculates the shortest distance, which separates it from a given destination, based on a full image of the network created using the most recent links from all routing nodes.

In the case of the "distance vector" method, each node transmits its vision of the distances separating it from other nodes in the network to its neighbors. Based on information received by its neighbors, each routing node then carries out a certain calculation to find the shortest path toward any given destination. The calculation process is repeated if the minimum distance between two nodes changes, until the network reaches a stable state. This technique is based on the distributed Bellman–Ford Algorithm (DBF) [GOL 93].

The drawback with protocols of this type is the cost involved in maintaining topological and routing information unnecessarily and in the absence of data traffic. This consideration means that these protocols are only used for relatively small networks. The control flow results in continuous bandwidth use, which increases rapidly as the number of nodes in the network increases. This becomes problematic for large-scale networks [VAS 14].

Protocols based on proactive principles include OLSR [CLA 03], DOLSR [ALS 11], TBRPF [OGI 02], DSDV [PER 01], FSR [PEI 00], Predictive-OLSR [ROS 13], ML-OLSR [ZHE 14] and COLSR [LI 12].

3.4.4. *Geographic protocols*

In a geographic routing protocol, each node should know its own geographic position, that of its neighbors, and that of the destination node. Neighborhood information is periodically updated by means of Hello messages exchanged between neighbors.

Each node obtains information concerning its own geographic position using a system such as GPS (Global Positioning System). Data are then sent toward the geographic position of the intended recipient. The neighbor closest to the destination retransmits the message. The difficulty here lies in the fact that each node needs to know the positions of all of its neighbors. Hello messages, sent out at regular intervals (proactive mechanism) or on demand (reactive mechanism) may be used to obtain approximate values. Using the reactive method, neighborhood requests are only sent out when a node needs to know the positions of its neighbors, whilst the proactive method floods the network in order to obtain the latest position of each node. Messages used to identify the positions of neighbors form the bulk of the overhead in geographic protocols. The time taken to send and execute this information also has repercussions for the transmission delay for messages between a source and a destination in the network.

Following a greedy geographic forwarding approach, the closest neighboring node to the destination is selected as the next hop. Proximity may be calculated as a Euclidean distance or using a linear projection between the current node and the destination.

Protocols based on this principle include Greedy Perimeter Stateless Routing for Wireless (GPSR) [KAR 00], Geographic Position Mobility Oriented Routing (GPMOR) [LIN 12a], Mobility Prediction Geographic Routing (MPGR) [LIN 12b], Location Aware Routing for Opportunistic Delay Tolerant Network (LAROD) [KUI 11] and Reactive-Greedy-Reactive (RGR) [SHI 12].

3.4.5. *UAANET networks and routing protocols: discussion*

Drone fleets allow multiple drones to coordinate their actions autonomously in order to improve the capabilities of the UAS system. Communication between drones operates through a communications network, which needs to guarantee that data will be delivered successfully whilst adapting dynamically to the implementation conditions of a mission. The ad hoc network structure presents a number of advantages in this context due to its reconfiguration capacities; routing protocols are used to manage drones entering and leaving the network. This architecture is also suited to the exchange of various real-time flows, as dynamic routing offers backup routes if links are broken or if obstacles are encountered between a source and a destination.

A dynamic routing protocol must be established in order to use a UAANET network, enabling communication between nodes. The choice of a UAANET routing algorithm is a complex task, as it needs to respond to the specific challenges posed by a UAANET network:

– total absence of a fixed infrastructure to distribute routing information;

– the topology of the network, characterized by low node density and a tendency to evolve over the course of a mission;

– the relative mobility of nodes. The Delair-Tech DT 18 drones [DRO 16] used in the context of this work operate at a speed of 17 m/sec;

– real-time application flows, as discussed above, which require low exchange latency;

– network integrity, that is, the need to reject fake packets or packets that have been modified by an attacker.

Considering these criteria, we see that proactive protocols are unlikely to be suited to low-density, high-mobility networks, such as UAANETs, as network capacity is only used to update routing information. The DSDV and OLSR protocols are particularly unsatisfactory for high-mobility UAANETs.

Furthermore, the low node density found in UAANETs means that there is little justification for using a hierarchical protocol. There are not enough nodes to form the node classes required when constructing a hierarchical topology.

In the case of geographic routing protocols, no concrete approach has yet been proposed for acquiring and sharing the coordinates of a destination node within a network; at the time of writing, only a few incomplete ideas have been put forward [MAU 01]. The first approach consists of using a central server for requests and responses regarding the geographical position of a destination node. The second approach consists of requesting localization information in a proactive or reactive manner. The main drawback of these approaches lies in the presence of a centralized server within the network, creating a need for an additional security mechanism as the server constitutes an added weak point in the final system. The time taken to execute requests and responses is also non-negligible. In the second case, the use of reactive or proactive position requests results in a lag, meaning that the obtained position may not correspond to the real position of the node.

Reactive protocols, on the other hand, are suitable for use in small-scale networks, where the delay involved in establishing routes is less significant. The use of on-demand route-finding ensures a rapid response if a link is lost. A reactive approach thus appears to be a good choice in our case, which only involves a limited quantity of routing packets. Nodes only generate route discovery packets as needed. Furthermore, whilst reactive protocols are not generally used for real-time applications (due to unpredictable delays), the low node density in UAANETs reduces the impact of the lag. Studies have shown similar performances in terms of end-to-end delay using AODV and OLSR [MAX 15, HYL 07]. This type of routing is therefore suitable for networks with dynamic topology. Experiments have shown that the AODV protocol offers good results in terms of transmission time and rates, whatever the level of node mobility [KHI 12]. TORA does not perform as well due to network overload.

However, it is hard to identify one protocol as being better than the others. Each routing protocol presents certain advantages in relation to density or mobility characteristics, specifications for the desired system or mission type. Moreover, the simulations presented in existing publications used hypotheses which were different from those involved in the real-world implementation of a drone fleet, touching on considerations such as mobility models, real-time requirements, or the assumption that the lower layers of the system behave perfectly. This situation is not representative of the real criteria which must be taken into account in UAANET networks.

3.5. Security in an ad hoc drone network

Work on UAANET networks has, to date, mostly focused on improving communications between drones and ground control stations and on optimizing communications between drones. This focus is due to the service quality requirements for UAANET networks, notably in terms of bandwidth usage by routing packets and the time taken to establish routes.

However, spontaneous environments of this type are subject to a number of security issues. First, given the use of wireless connections, the network is intrinsically vulnerable to illicit listening or service refusal attacks. Secondly, the absence of a centralized element means that there is no message and node authentication service. This means that fake packets may be injected into the network, disrupting the sequencing activities of the routing protocol. This type of attack may corrupt communications or reduce network performance [BUC 02]. Security mechanisms designed to identify these weaknesses must take account of limitations in terms of energy, memory and bandwidth. These constraints necessitate an approach based on preventive security mechanisms, with high resistance to the presence of compromised nodes in the network. Given the sensitive applications for which UAS are used, attackers must be prevented from accessing the application flows, and thus corrupting the viability of the UAS system [AKR 14].

In this section, we shall discuss security elements relating to ad hoc drone networks. We shall begin by summarizing causes of weakness in UAANET networks, analyzing the risks in order to identify the most probable types of attack which may be directed against the network. Next, we will specify the security needs of these networks, highlighting the functions and data within the network which need to be protected. In the final section, we shall present existing security solutions used in the MANET family of networks, before studying their application to UAANETs. Lastly, we will indicate a number of promising areas for development in order to improve the security of routing traffic.

3.5.1. *Weaknesses in UAANET networks*

Causes of weakness in UAANET networks include a number of elements which are intrinsic to the UAS system environment [KIM 12]; that is, which are linked to the physical architecture of nodes, communication links, and the conditions of implementation of the UAS system (such as the need for cooperation between nodes). A second source of weakness in UAANETs relates to the potential presence of malicious nodes in the network [JAV 12]. These different weak points are analyzed below.

1) *Vulnerable communications channels*: wireless connections are used to send and receive signals in a UAANET network. These radio links may be subject to a variety of attack types, including illicit listening or active interference [KAN 07]. Listening attacks on ad hoc wireless networks generally consist of placing a malicious node between two or more communicating nodes. Given that all traffic is airborne, the attacker simply needs to adopt a position in the zone covered by the target nodes in order to intercept traffic. If the aim is to listen to the whole network, an attacker may also act on the routing protocol, generating false packets or modifying routing packets, ensuring that all traffic is diverted via the attacker. In the specific context of UAANET networks, an attacker using a high-gain antenna within the range of a drone may be able to listen to all communications within the fleet. Furthermore, depending on the characteristics of data connections,

links often have low bandwidth. Attackers can exploit these characteristics by transmitting packets in order to saturate the network and thus block communications between nodes [YI 05].

2) *Uncontrolled environment*: wireless ad hoc networks operate in a distributed and dynamic environment. This means that communications between nodes participating in the routing process are shared and operate in an opportunistic manner. This makes it hard to control nodes coming into or leaving the network, and a malicious node may be able to connect to the network and participate in packet transfer. An attacker may also usurp the identity of a genuine node, falsifying the routing mechanism by emitting erroneous information or retransmitting out-of-date information. Rushing attacks operate in this way [HU 03]. UAANET networks, like other ad hoc networks, thus have no inherent protection against intruders.

3) *Dynamic topology*: the high travel speed of UAVs (DT 18 drones have a maximum speed of 80 km/h) means that the topology of the network is not stable and changes continuously (in response to C2 traffic sent by the GCS and to the drones' own flight plans). This characteristic creates security issues, as routing protocols do not have the intrinsic ability to differentiate between interrupted communications (or a broken link) caused by drone movements and those caused by an attacker acting on the network [ELS 09]. Both situations can cause a group of nodes to become disconnected from the network; the representations at application level are also identical. Moreover, if the malicious node cooperates partially in transferring information whilst creating incoherences in the routing protocol, it cannot be detected without the use of more sophisticated network analysis tools applying reputation and confidence metrics [SEN 10].

4) *Cooperation problems*: in an ad hoc wireless network, routing algorithms do not intrinsically require the use of a node pre-association algorithm prior to execution. This is based on the hypothesis that nodes are cooperative and not malicious; situations involving a malicious node compromising the network are not taken into account. Consequently, node authenticity is not guaranteed. It is therefore easy for malicious nodes to enter the network, publishing routes with better metrics.

5) *Limited resources*: the calculation and memory capacities of drones are limited by their size. This fact may be exploited to attack the network by exhausting drone resources. The sleep deprivation attack [PIR 06], for

example, consists of transmitting a constant stream of control messages toward network nodes. Once all resources have been used up, drones may be captured or diverted by an attacker. Furthermore, the introduction of security mechanisms in a network can also reduce performance by increasing demand on the microprocessors [YAN 04]. A compromise needs to be found between the reliability of the communications network and security.

6) *Existence of attacks*: ad hoc networks are generally targeted by attacks directed against a sub-set of the protocol layer of the OSI model [FAD 10]. These attacks may be grouped into two classes: rational and irrational. Irrational attacks aim to disrupt the operation of network services without deriving any benefit from the results, whilst rational attacks are designed to provide a direct or indirect benefit for the attacker. For example, they may aim to violate a network security policy in order to falsify functions and sensitive data. Sensitive data in a network may relate to elements such as routing, energy management configuration or security (cryptographic keys, certificates, etc.). Violating the integrity of this critical information may lead to a variation in network connectivity, causing the mission to fail.

3.5.2. *Attacks on UAANET networks*

To model an attacker, we need to quantify their capacities. In general terms, the more resources an attacker possesses, the harder it will be to resist the attack. Other characteristics which must be taken into account when designing security mechanisms include the number of attackers, their level of coordination and their technical capacity. Several attackers may attempt to disrupt a network at the same time, acting autonomously or simultaneously in order to increase their chances of success.

Additionally, an estimation of the technical capabilities of attackers is important in identifying the nature of threats. For example, a malicious node may only be able to receive data, but not create false packets. An attacker may also present as a valid node and be able to access the full range of network services. A malicious node may be able to access the whole network by diffusing false packets, or only part of the network.

We also need to consider the area targeted by the attack. If an attacker targets a low-level layer which affects the layers above it (the link layer, the network layer and so on), several network services, such as localization, time syncing or energy management may be disrupted. In a UAANET network, critical services require a higher level of defense than optional services, as a malfunction in a critical service may disrupt the whole network.

3.5.2.1. *Attacks on the physical layer*

Attacks targeting the physical layer of a system act on hardware (such as the wireless modem). This type of attack may be carried out with no knowledge of network topology or of the technology used by the nodes [GOK 16]. From the attacker's perspective, this type of attack requires the use of specific materials [WU 07] with the capacity to disrupt exchanged signals (e.g. a high-gain antenna).

Attacks on this layer typically include illicit listening, active interference and jamming attacks. The attacks are made possible by the fact that the mobile nodes share the same wireless medium. This makes it easy to pick up transmitted messages and to inject fake messages into the network. Note, however, that attacks of this kind are considerably harder to carry out on high-mobility networks [WU 07]; an attacker must stay within range of their target node to have an effect on communications. As we saw in Chapter 1, in the specific case of UAANET networks, drones move in a dynamic manner due to the succession of commands sent by the GCS or their own preloaded flight plan. This type of mobility pattern makes it hard for attackers to get close to their targets. For this reason, we have decided not to consider this attack in our study of secure communications architectures.

3.5.2.2. *Attacks on the link layer*

The protocols in the link layer are designed to coordinate transmission over the common medium. This layer is notably responsible for communications between neighbors. The widely used IEEE 802.11 protocol [OHA 05] does this by means of distributed contention management algorithms to divide up the medium. There are two types of DCF or PCF algorithm, described in detail in [ZHU 03].

Routing and MAC protocols generally work on the hypothesis that each entity respects the specifications of the protocol and that entities will cooperate. However, in the presence of a dynamic topology, this hypothesis ceases to be realistic, as there is nothing to prevent a malicious node from compromising the network at different levels of the protocol layers of the OSI model. The IEEE 802.11 layer is therefore potentially vulnerable, as there is no central entity to monitor communications between participants. A wide variety of attack types target this level. They may work by increasing the time between frames (Shortest InterFrame Space: SIFS; Clear to Send: CTS, DCF; and Polls IFS: PFS) to a value above the threshold of acceptability [KYA 05]. Another technique is to exhaust the resources or the capacity of the transmission channel via a service refusal attack [RAY 08] or by preventing access to the necessary travel [RAY 04]. A number of different approaches for detecting service refusal attacks have been put forward [SU 08, SHA 11]. Most focus on detecting backoff manipulation by applying confidence measurement mechanisms between nodes, whereby each node monitors the value of its neighbor's backoff variable. This results in the creation of a chain of trust, which is used to evaluate variations in network performance during communications [GUA 05]. However, this technique has certain limitations. One issue concerns the hypothesis that authentication and integrity security mechanisms are already in place in the network [YAN 04]; the development of these algorithms is a significant issue in its own right.

Techniques for resisting attacks on the MAC level are generally based on the use of chains of trust between nodes. However, for an algorithm of this type to operate, we must suppose the existence of a reliable and secure routing protocol, capable of authentication and of verifying message integrity. This means that security propositions for the MAC layer cannot be treated within the link layer alone. To the best of our knowledge, all of the solutions currently available involve a solution at link level which must be used in association with a solution at network level. Consequently, we may decide to only secure the network level, adding blocks to secure the link level at a later point. We may also choose to treat the two areas together, taking a cross layer approach such as that described in [SU 08].

We shall use this first principle in what follows, focusing exclusively on proposing an authentication and integrity verification mechanism at network level. This approach also allows us to concentrate on protecting multi-hop communications. The proposals put forward in [BAR 16] may be used to secure communications at link level, but will not be discussed further here.

3.5.2.3. *Attacks linked to routing protocols*

The network layer allows several nodes to connect and form an ad hoc network. This connectivity is made possible by collaboration between all of the nodes in the network. However, the collaboration hypothesis is hard to maintain in a dynamic environment with nodes fulfilling different roles. A range of different attacks therefore target the network layer; these have been identified and studied in a number of publications [KAN 07]. The aims of these attacks vary. They may be intended to absorb network traffic, to insert malicious nodes into the network, or to divert and control network flows. Several methods may be used to these ends. Network traffic may be re-transmitted along a non-optimal route, creating significant delays in comparison with the normal communication time. Packets may also be re-transmitted down non-existent routes, resulting in their loss. Attackers may create routing loops to create full or partial blockage of the network, or the network may be partitioned by multiple attackers, preventing a source node from finding a route to its desired destination and thus degrading network performance [KHO 08]. Some examples of attacks operating on routing protocols are presented below.

Black hole attacks: in a black hole attack [TAM 07], one or more attackers form a zone within a network, in which traffic is subtly made to disappear without informing the source node. Depending on network density, this attack may be carried out by one or more malicious nodes acting independently (single black hole) or together (collaborative black hole) [ARA 16]. The principle is to convince the valid nodes that there is a better route to a given destination, for example, by inserting fake control packets into the network. If these packets are processed, then a route will be created through a black hole. Another variation of this attack is the grey hole method [KAU 14]. In this case, only some, not all, packets are deleted.

This type of attack is illustrated in Figure 3.4. In this example, the ground station is attempting to establish a route toward node Dr5. The black hole attacker is able to usurp the identity of a valid node in the network, and can thus participate in the route discovery mechanism, sending a route reply message back to the GCS. This message announces a route to node DR5 with a minimal cost (e.g. with a lower hop count than other routes). The GCS updates its routing table to include this fake route. Data packets will then travel through the malicious node, which simply captures and ignores them.

Wormhole attacks: wormhole attacks [MAH 08] generally consist of placing a tunnel between two attackers (acting in cooperation) in the network. More than two attackers may be present, however, creating multiple tunnels as a function of node density. In this first configuration, the first attacker carries out illicit listening activities in the zone covered by a target node. It then retransmits these packets through the tunnel to the second attacker, who may then re-inject them into the network, with or without modifications. If the attackers modify the contents of these packets, they will be able to modify the topology of the network unless a message authentication mechanism is present. However, attacks of this type are generally executed without packet modifications in order to circumvent authentication algorithms. Note that whilst it is effective against all types of routing protocols, the wormhole attack type is particularly effective against protocols which use hop counts as their routing metric. Once a route containing malicious nodes has been selected, attackers may act in other ways by modifying or selecting packets. This type of attack disrupts packet routing, as distant nodes may believe themselves to be neighbors. Aiming to choose the shortest route, genuine nodes will select the wormhole tunnel by default rather than taking a longer, legitimate route. In this way, attackers are able to short circuit normal routing message emissions, creating a virtual tunnel and splitting the network. A wormhole attack can only be detected by identifying genuine packets exchanged through a wormhole connection, something which is difficult to do in the case of a mobile ad hoc network.

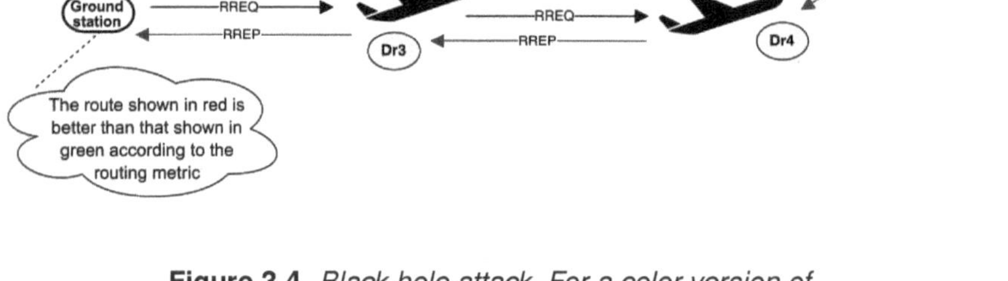

Figure 3.4. *Black hole attack. For a color version of this figure, see www.iste.co.uk/maxa/software.zip*

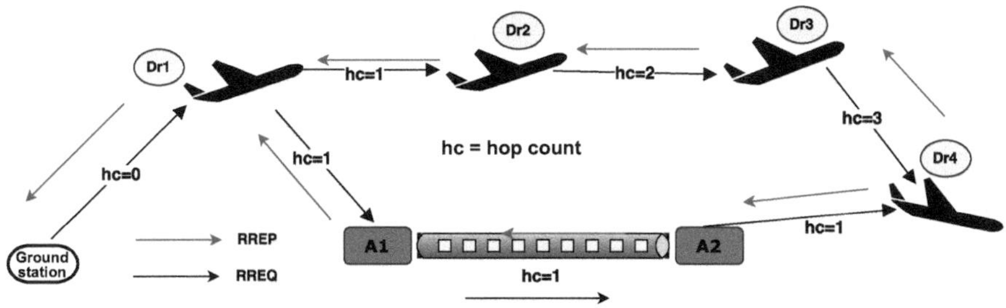

Figure 3.5. *Illustration of a wormhole attack. For a color version of this figure, see www.iste.co.uk/maxa/software.zip*

Figure 3.5 shows an example of a wormhole attack in which S and N3 are tricked into believing themselves to be neighbors. In this case, A1 listens to node S to receive and transfer packets directly to A2.

This leads N3 to believe that S is within the range. Control packets (i.e. Hello messages, RREQ requests, RREP responses and RERR error messages) are thus transmitted directly from node S to N2. Even without any knowledge of the cryptographic keys or hashing function used by the system, attackers are able to reduce network integrity by diverting control packets. In the wake of this attack, hostile elements may degrade network performance by selectively filtering packets, or by carrying out a more complex attack, such as rushing.

The following section is devoted to existing security approaches used to combat this type of attack. The SAODV routing protocol [ZAP 02a] will be used to illustrate the limitations of authentication services when faced with a wormhole attack. A full security solution for our planned UAANET network will then be presented.

3.5.2.4. *Existing solutions and their limitations*

The security requirements of UAANET networks are similar to those of other types of MANET network. However, differences exist in terms of the real-time and critical character of application flows. Note, too, that any implementation of security mechanisms to take account of a sub-set of these security services must also take account of application requirements, particularly in terms of time, reliability and bandwidth optimization. This means that the security layer should not adversely affect network performance. Processing times must be kept short in order to avoid impacting the transmission delay. Energy use also needs to be optimized in order to avoid connectivity losses. The security services needed for a UAANET network to function correctly may be defined as follows:

– Authentication: used to check whether or not the identity of a node in a network is genuine. This service is important during the route discovery phase in order to authenticate nodes and the messages they exchange. In the absence of an authentication mechanism, adversaries are able to fake routing control information, usurping the identity of a genuine node and thus intercepting application flows. For example, a malicious node may take the place of the GCS in order to receive all payload traffic.

– Integrity: this consists of verifying whether or not data packets have been tampered with during transfer from a source to a destination. It ensures that data have not been modified in a way which might change the nodes' vision of network topology. During an attack targeting system integrity, a malicious node may attempt to convince a non-neighboring node that they are, in fact, neighbors. Integrity is verified using mutable fields, that is, fields that change within the network.

– Confidentiality: in the case of UAANET networks, this property ensures that real-time and critical data can only be accessed by authorized nodes. It is important to note that, in general, confidentiality is applied to application traffic rather than to routing traffic. Routing traffic does not contain any sensitive information and thus does not need to be confidential.

– Availability: this element guarantees that all network services within a UAS system are available. Availability is hard to guarantee due to the specific constraints of UAANET networks. Generally speaking, it is impossible to maintain a network state for a MANET network in the same way as for a wired network. The topology and state of these networks evolve over time due to node movements.

3.5.3. *SAODV secure ad hoc routing protocols*

The SAODV protocol is a secure routing algorithm that proposes methods to prevent malicious nodes from joining ad hoc networks. It is important to note that several types of secure protocols have been proposed in the published literature, most of which are based on the OLSR, AODV and DSR protocols (see Table 3.4). In our work, for reasons of clarity, we have chosen to concentrate on the SAODV protocol, as this option presents significant differences from other existing algorithms. The reasons for our choice will be discussed in the following section.

Protocol	Security service	Security mechanisms	Advantages	Disadvantages
SAODV	– Authentication – Integrity – Non-repudiation	– Hash chain – Digital signature	– Secures different control packets: RREQ, RREP, RERR, Hello – Robust against modification-type attacks (black hole, grey hole and sink hole) – Secures mutable and non-mutable fields	– Vulnerable to wormhole and rushing attacks
SRP	– Authentication – Integrity	– HMAC	– Security against modifications and fake messages	– Does not secure route maintenance packets – Vulnerable to wormhole and rushing attacks
ARAN	– Authentication	– Certificate server	– Against spoofing, replay and routing message modification attacks	– Based on the use of a centralized server – Not possible in a UAANET – Vulnerable to wormhole and rushing attacks.
ARIADNE	– Authentication – Integrity	– Hash chain – TESLA – Digital signature	– Authentication of nodes along route – Against modification type attacks	– Synchronization using TESLA may not be reliable in a UAANET – Vulnerable to wormhole and rushing attacks
SOLSR	– Authentication – Integrity	– HMAC	– Against replay attacks	– Increases overhead in cases of high mobility. – Incoherency between node position and information contained in packet.
SEAR	– Authentication – Integrity	– Hash chain	– Security against modifications and fake messages	– Vulnerable to wormhole attacks
SEAD	– Authentication	– One-way hash function	– Against service denial and resource exhaustion attacks	– Vulnerable in cases with multiple assailants: e.g. wormhole attacks, collaborative black hole attacks
SPREAD	– Authentication – Integrity – Confidentiality	– Confidence metric – Multi-path routing	– If the source and the destination are secure, SPREAD is able to secure data – Data are shared through multiple connections	– Whole network is vulnerable if the destination is attacked. – Vulnerable to black hole, wormhole and rushing attacks

Table 3.4. *Summary of existing secure routing protocols*

The SAODV (Secure AODV) protocol [ZAP 02a] is designed to guarantee message authentication. Non-mutable fields of control packets (such as the source and destination IP addresses) are successively signed by each intermediary node, and these signatures are authenticated by each subsequent node. Mutable fields (such as the hop count), considered to be more sensitive, cannot be signature-verified due to the increase in processing time which this would entail. The concept of network load also needs to be taken into account, as checking signatures for each retransmission adds to the security overhead. Consequently, the hash chain principle [HU 04b] is used to ensure that hop counts have not been modified by an attacker.

Note that the SAODV protocol is used in conjunction with a key management infrastructure to manage the lifecycle of keys between the nodes in a network. Using the message signature algorithm, nodes must use a pair of keys for asymmetric encryption. Network nodes also need to have access to the public key for each other node in the network.

The SAODV protocol offers a robust defense against attacks aiming to modify genuine packets and/or to create new packets which do not conform to the relevant routing rules. It is also able to withstand identity theft attacks. The main aim of SAODV is to ensure that malicious nodes cannot usurp the identity of genuine nodes and interact with the network in the place of the victim node. Intermediary nodes must also be prevented from modifying packets in transit. This property constitutes a means of defense against a certain number of attacks, including the black hole attack described above.

However, the protocol does not offer total protection; for example, an attacker might still be able to increase the hop count. The value of the hash field remains unchanged in these cases, meaning that genuine nodes will not be able to detect malicious activity of this type, including wormhole attacks.

In the following section, we shall analyze the SAODV protocol in order to provide a formal description of its shortfall, before proposing our own original secure routing protocol.

3.5.3.1. *Secure routing protocols currently available*

Table 3.4 provides a summary and comparison of existing secure routing protocols. The following parameters should be taken into consideration when choosing a protocol.

– Suitability for dynamic topologies: as we have seen, UAANET networks are characterized by a rapidly changing topology, resulting from the speed of movement of the drones involved. It is therefore best to select a protocol that adapts well to different mobility scenarios, and which is able to maintain acceptable performance (in terms of packet loss rate, transmission times, route re-creation and overheads) under these conditions.

– Overhead: this relates to the bandwidth occupied by the header of the security element of a routing protocol. The overhead generated in this way should not have a negative impact on the application flow.

– Memory and processing capacity: this metric is used to associate the chosen approach with its requirements in terms of memory and processing capacity. The chosen approach should use as few resources as possible.

The metrics listed above should be taken into account when comparing ad hoc routing security protocols for UAANET networks. We need to respond to the following questions:

1) Does the security objective of the UAANET correspond to that of existing secure routing protocols?

2) Are the security mechanisms proposed for MANET networks suitable for UAANETs?

3) What are the specific challenges presented by the UAANET network, which are not covered by existing ad hoc security mechanisms?

Considering existing approaches, we see that the security objectives for UAANETs are the same as those for MANETSs. We aim to guarantee the authentication of routing messages, the integrity of mutable fields and the authentication of nodes; this latter element is covered by an adaptive key management infrastructure. In this case, we have chosen to ignore the question of confidentiality for routing messages, as routing information does not contain sensitive elements which could be used to disrupt the network. Only message authentication is truly essential for detecting unauthorized modifications to routing packets or the creation of fake packets. For this reason, we can also ignore the "secrecy and anonymity" security services used in certain VANET networks [ENG 14].

Furthermore, due to the high mobility of drones and the changeability of network topology, we shall not consider algorithms that present scalability problems, such as SOLSR and SRP. These protocols generate a significant quantity of information packets for each node to exchange or maintain, something which is problematic in a dynamic topology network such as a UAANET, where the delays and the scale of the overhead involved can have a negative effect on network performance.

It is also important to note that the scalability metric is not particularly significant in a UAANET network, mostly because of the low node density. On this point, symmetric-key cryptographic techniques appear to be particularly appropriate. However, key sharing approaches become complex very quickly as the number of nodes in the network increases. This problem may be attenuated using a diffusion-based authentication architecture such as TESLA; however, the problem may remain due to the real-time nature of application traffic, which the TESLA algorithm is not always able to guarantee. Using TESLA, authentication creates a certain delay in each direction due to the key acquisition procedure. Another reason not to use symmetric-key cryptography relates to the fact that UAANET nodes do not cooperate with a central entity for identifying and eliminating malicious nodes. A non-repudiation service is necessary in order to resolve these problems, and this can only be obtained using asymmetric (public-key) cryptography. Note that public-key and symmetric-key cryptography may be combined for reasons of reliability. However, as we have seen, a purely symmetric approach is not suitable for UAANET networks, and we may eliminate the SRP, SEAR and SAR protocols from the list of possibilities.

The major drawback to public-key approaches lies in their processing and memory requirements, which may be too much for drones with limited memory and energy capacities [SAL 13].

In this specific case, the drones used have far higher memory and processing capacities than those of mobile nodes in MANET networks. For this reason, the problems associated with resource usage for public-key cryptography in MANET networks may safely be ignored in our study of security protocols for UAANET networks.

We may also eliminate those solutions requiring the use of a certificate server, such as the ARAN protocol. Systems of this type are not yet available for the architecture in question [GUP 15]; furthermore, they can create additional weaknesses in cases of attack on a centralized system.

The SPREAD protocol sends some of its traffic across multiple connections, which means that several routes, and thus several nodes, are needed. UAANET networks often only involve three or four drones, so multiple routes between a source and a destination are rare.

Based on these considerations, the SAODV appears best suited to our needs. It guarantees authentication and message integrity, whilst providing security for both mutable and non-mutable fields. Mutable fields are protected by a hash function in order to reduce the processing power requirements for each hop. Our UAS system has the energy resources necessary for public-key cryptography. The elements involved in our choice are summarized in Table 3.5.

Technical solutions	Algorithms	Security services	Offering resistance against
Digital signature (SAODV)	RSA	Authentication	– Performance degradation – Modification of network topology
Hashing function (SAODV)	SHA [GIL 04]	Integrity	

Table 3.5. *Selected mechanisms*

Note, however, that the SAODV protocol is vulnerable to different variations in the wormhole attack, notably when one or more attackers transfer routing packets without modifying the hop count. The SAODV protocol is not able to detect these attacks. We have therefore considered this type of attack specifically in order to offer a solution, discussed in the following section.

In the context of this study, we have also considered the means of contributing to the certification of our secure communications architecture. This certification is necessary if our UAS system, with its ad hoc communications architecture, is to be used in national air space. Delair-Tech, the company which manufactures the drones used for this study, has also expressed this requirement for validation of the UAS system involving our ad hoc communications architecture. There is currently no official standard governing the deployment of drone fleets; for this reason, we have considered standards applicable to other areas of the aeronautical sector, notably the

DO-178 Software Considerations in Airborne Systems and Equipment Certification [BAG 07]. In this sector, embedded software for aircraft must respect certification constraints by ensuring that the development process conforms to standard requirements, following, for example, a specific task sequence (following the V cycle). In this study, we have considered the practical implementation of a rapid prototyping methodology for software development (based on a model-driven development, or MDD, approach) in relation to our secure routing protocol.

3.6. Designing a new secure routing protocol for UAANETs (SUAP: Secure UAANET Routing Protocol)

In this section, we shall propose a new routing protocol, designed to respond to the reliability and security requirements for communications within a drone fleet.

In creating our protocol, we began by studying a number of routing protocols found in the existing literature [MAX 17]. We noted that none of these existing algorithms include security mechanisms with the ability to verify the authentication of identified routes. Systems are subject to different types of attacks. These attacks consist of modifying routing packets, or injecting fake messages into the system with the aim of degrading network performance. A secure communication architecture, able to take account of the sensitivity of potential applications, is therefore of critical importance. Routing security is a particularly challenging issue due to network specificities in terms of mobility, connectivity and available resources (energy, memory, bandwidth, etc.). Furthermore, most existing secure MANET routing protocols are vulnerable to wormhole attacks, which are one of the best-known threats to ad hoc routing.

In response to this problem, we have developed a secure routing protocol for UAANET networks, SUAP (Secure UAV Ad Hoc Routing Protocol [MAX 16a, MAX 16b]). This protocol is based on the SAODV protocol, and provides security for the route discovery and maintenance stage. SUAP guarantees the authentication and integrity of control packets via a signature mechanism and a hashing function. Additional security mechanisms were then added to detect wormhole attacks. The first element of this mechanism is devoted to neighbor discovery packers, and consists of establishing a correspondence between the relative distance between two neighboring nodes and the hop count associated with a packet. The second element protects route discovery packets using the identity of the following node in the route chain to calculate a "footprint" (hash) for each retransmission.

We shall begin our presentation of SUAP by discussing the hypotheses used in relation to the network model and our chosen attack model. We shall analyze the choices made at the start of the process in constructing the basis for the SUAP protocol. Next, we shall describe the different security mechanisms used in the SUAP protocol, highlighting the vulnerabilities which we considered during the design phase. We shall then analyze the formal verification process used to verify the security services provided by SUAP using the AVISPA tool [ARM 05]. Finally, we shall provide a detailed description of the design process used to create the SUAP protocol using the rapid prototyping methodology described in the previous chapter.

3.6.1. *Choosing an initial routing protocol*

In constructing our secure communications architecture, we needed to choose an existing routing protocol as a starting point. We selected a protocol used in the MANET environment, adding our own security mechanisms. This choice is due, on the one hand, to the maturity of MANET routing protocols (notably AODV, DSR and OLSR), discussed at length in a range of literature, including performance evaluations [MAC 99, PAU 12] and, on the other hand, to the existence of standards in this area. One example includes the RFC[3] used in the AODV [PER 03b], DSR [JOH 07] and OLSR protocols [CLA 03].

In order to select a protocol, we needed to evaluate the main families of routing protocol and to identify those which come closest to the real implementation conditions of a UAANET network. For this reason, we needed to consider the execution environment of routing algorithms to find the closest match. In this context, the conditions of implementation relate to the Linux operating system used by the drones, real-time execution of routing protocols and realistic mobility models. To respond to these requirements, we used an experimental protocol to evaluate routing protocols following clearly defined parameters specific to the UAANET network. For this reason, we

3 Request For Comment

created a specific tool to test UAANET routing protocols, combining both simulation and emulation paradigms. This protocol comparison tool may also be used for other network types [MAX 15], such as VANET or AANET (Aeronautical Ad Hoc Network), with the addition of the appropriate modules.

The details of our evaluation process are set out in [MAX 15]. Our results showed that the AODV protocol performs better in these conditions than OLSR or DSR, given our specific scenario and considering metrics including the connectivity rate, end-to-end delay, average re-creation delay and overhead rate.

We therefore chose to use the AODV protocol as the basis for our secure routing protocol. The mechanisms in the SUAP protocol which relate directly to route creation in a network are based on AODV. In section 3.8, we shall present the model-driven implementation of this new protocol, using the rapid prototyping methodology described in the previous chapter.

3.6.2. *The SUAP protocol*

3.6.2.1. *Network model and attacks considered in designing the SUAP protocol*

In this section, we shall consider the hypotheses used in designing the SUAP protocol.

– As in any type of MANET network, the routing protocol carries two types of information: routing packets and information generated by the payload (e.g. video recorded by a drone). These two types of message are of differing natures and have different security requirements. In this study, we have chosen to focus on the security of control packets linked to the routing protocol. These packets are generally transmitted between neighbors, and may thus encounter a malicious node on the route between the sender and the receiver.

– Nodes require access to a public-key infrastructure to generate and revoke cryptographic keys. Services provided by PKIs involve the creation, management, storage and revocation of certificates.

– We considered the nodes in our network to be homogeneous; the drones and ground stations in our study were all made by the same company (Delair-Tech). This hypothesis allows us to ignore potential weaknesses caused by heterogeneity in the network. The publications cited in [ERM 08] and [LI 09] stipulate that the use of a heterogeneous group of nodes may result in security problems due to "selfish" node behaviors.

– We also considered that the nodes in the network had access to sufficient resources to apply cryptographic algorithms such as RSA-512 and SHA-256. DT 18 drones have enough processing power to justify this hypothesis, allowing us to apply algorithms which are considered to be too heavy or resource-hungry (in terms of memory and energy) for other types of ad hoc network [PER 02, MOH 09] (e.g. probe networks).

– Furthermore, note that all nodes in the network are presumed to be equipped with an omnidirectional antenna. The communications range of each node is limited, so the distance between nodes can be represented by the variable $r < D_{max}$, where D_{max} is the maximum range distance of a node.

– We presumed that all nodes in the network were time-synchronized, something which is made possible by the use of GPS equipment in the drones. This hypothesis is taken into account for our geographical leash algorithm, described in section 3.6.7. The possibility of synchronizing drones is specific to UAANET networks, and is often difficult to achieve in MANET networks [JAI 09].

– Our attacker model is based on the Dolev-Yao model [CER 01]. The Dolev-Yao model presumes that intruders are able to intercept encrypted messages, but cannot modify them without the corresponding security key. As we stated previously, we have chosen not to consider attacks on the encryption method, and our approach includes the hypothesis of perfect encryption. Thus, a message cannot be decrypted without the corresponding key. Intruders are only able to access the public keys of all nodes, their own private key, agent identities (IP addresses) and public data contained within the protocol, where applicable.

3.6.2.2. *Description of the SUAP protocol*

The SUAP protocol [MAX 16a] is a reactive secure protocol, designed for execution in ad hoc drone networks. It is based on the AODV [PER 03a] and SAODV [ZAP 02a] protocols. SUAP is characterized by security mechanisms that guarantee the authentication of non-mutable fields (i.e. fields within the routing protocol which remain static from emission to reception of a packet by the receiver, such as the addresses of the source and destination nodes), the integrity of mutable fields (such as the hop count) and non-repudiation. It also includes mechanisms for detecting wormhole attacks.

In what follows, we shall begin by considering the mechanisms and vulnerabilities encountered in the SAODV protocol, studying the execution of the wormhole attack in a UAANET network. We shall also highlight a particular vulnerability of the SAODV protocol to attacks in which a malicious node listens to the network, transfers a packet, but does not modify the packet in order to avoid detection by the authentication mechanism. This attack (similar to the wormhole attack, but carried out by a single node) is intended to convince two distant nodes that they are neighbors. Secondly, we shall provide a detailed analysis of existing contributions, focusing on ways to combat attacks of this type and highlighting the drawbacks of current approaches. Finally, at the end of this section, we shall discuss the specification of the SUAP protocol.

3.6.2.3. *Analysis of existing solutions*

A number of authors have studied the security of direct communications between GCS and drones. The aim in this type of architecture is to secure the physical modules of a system and to limit access to the communications link. As this type of configuration is very different from that used in our architecture, the solutions which have been developed for this context will not be considered here. In the specific case of drone fleets, the contribution which comes closest to our approach here is by Arkam *et al.*, published in [AKR 14], and relates to improving drone fleet security through the use of an embedded physical module in each node, providing a certain number of security services. In our work, we decided from the outset to focus exclusively on software solutions, taking account of the existing physical architecture of a UAS system made up of DT18 drones. We therefore turned our attention to the handful of secure routing protocols that have been proposed for MANET networks, finally opting to study the SAODV protocol, for a number of reasons:

– The protocol provides authentication services and ensures the integrity of routing messages, except in the case of certain security failures, which will be described in the next section. SAODV defends against attacks using a signature and hashing procedure to verify authenticity.

– The SAODV protocol provides the means of counteracting a black hole attack, ensuring that nodes from outside the network will not be able to transmit false packets following a route request. The SAODV protocol takes account of two situations in relation to route requests: either the destination node must respond directly to the request in order to prevent malicious nodes from creating fake routes, or an intermediary node may be allowed to respond, but only if it has previously been authenticated.

– The SAODV protocol also guarantees the integrity of the hop count for each retransmission. This service is important to prevent modification by malicious nodes, and is provided using a one-way hashing function.

– SAODV is not particularly costly in comparison with other existing security protocols (studied above), as routing messages are not signed with each retransmission. Only the source and destination nodes sign messages. Intermediary nodes simply verify the signature for each hop. It is thus appropriate to differentiate between approaches to mutable and non-mutable fields.

3.6.3. *The SAODV protocol*

The SAODV protocol is based around the use of a signature to protect static data in control messages via the use of a public-key encryption algorithm, whilst using a hash chain to protect the integrity of variable elements, including the hop count.

SAODV uses a double signature to authenticate response packets sent specifically by neighbors which know the destination[4].

3.6.3.1. *Digital signatures in SAODV*

The first security mechanism used in SAODV is the digital signature, a form of public-key encryption used to guarantee the authenticity of non-mutable fields. The SAODV protocol uses the fields from the AODV protocol, acting as an extension (see Figure 3.6 for an illustration of the SAODV extension format). A node emitting an RREQ or RREP packet, wishing to contact another node, signs each message it transmits; this prevents malicious external nodes from participating. On receiving these messages, intermediary nodes begin by verifying message authenticity before processing the packet.

---

4 Note that conceptually, in our view, the double signature principle adds nothing in security terms. We might consider deactivating the Destination Only field in the request packet to enable any neighboring node to respond if it is in possession of a route to the destination.

← 4 Octets →

Type (1 octet)	Length (1 octet)	Hash function (1 octet)	Max hop count (1 octet)
Top hash			
Signature			
Hash			

Figure 3.6. *Extension format of an SAODV message*

It is also important to note that route discovery packets may take two formats, one including a single signature, and the other with a double signature. Use of the double signature is justified by the fact that a malicious node might respond to a request by sending a fake RRES packet. In the AODV protocol, an intermediary node may respond to a route request if it knows a route to the desired destination. However, this chain cannot operate with a single signature approach, as the intermediary node is unable to sign the RREP packet in the place of the destination node. This issue is solved by the use of the double signature principle, which operates as follows:

Consider the configuration shown in Figure 3.7. When node S sends a request packet, it includes a second signature (see Figure 3.8) S. For instance, suppose that, in state t0, node S is seeking a route to a destination node D. In this case, an RREQ packet is sent and retransmitted up to node D. Next, suppose that node D wishes to find a route to node S; when the RREQ packet from node D arrives at node N2 (which we shall assume to be connected to node S through N1), this node may respond in place of S by sending an RREP to D. This algorithm is fully operational in a MANET context, but a few points should be noted with regard to its application in our network model.

– The use of a second signature is only justified if the nodes are heterogeneous, may be malicious, or may act selfishly. For example, a node may create fake packets to avoid being included in data routing activities. As the nodes in our network are presumed to be homogeneous, the double signature principle is not relevant.

– The second signature is only necessary if we are unable to guarantee authentication between neighbors. In Figure 3.7, for example, node D may not trust N3, as the latter node may be new to the network. In our routing protocol, SUAP is equipped with a neighbor authentication algorithm which is executed during the neighborhood discovery phase via an exchange of Hello messages. As we shall see, this procedure guarantees neighbor reliability; hence, a network node may allow its neighbor to respond in its place without requiring a double signature.

Figure 3.7. *Illustration of the hash chain mechanism in SAODV*

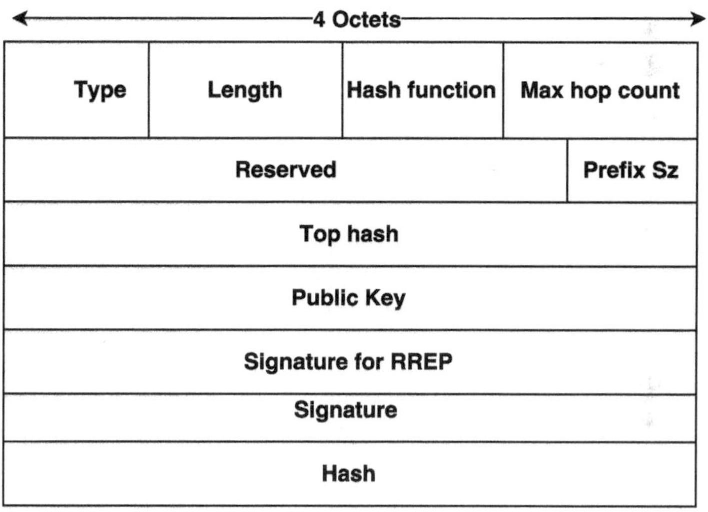

Figure 3.8. *Extension format of an RREQ message with double signature*

The extension format for the request packet with a double signature (as shown in Figure 3.8) is explained below.

– Type: indicates the packet type (equal to 64).

– Length: the size of the packet in bytes, excluding the *Type* and *Length* fields.

– Hash function: indicates the hash function used to calculate the *Hash* and *Top hash* values.

– Max Hop Count: the maximum number of hops, which may be set at 64 (TTL value) - (*network lifetime*).

– Top hash: a variable used to authenticate the integrity of the hop count.

– Signature: indicates the signature method used.

– Reserved: set at zero.

– Prefix Size: the size of the *Prefix* field of the RREP packet.

– Signature for RREP: the signature for the RREP packet, used to allow an intermediary node receiving the RREQ to respond to this message.

– Signature: the signature of all non-mutable fields.

– Hash: the value of the stamp corresponding to the current hop count.

3.6.3.2. *Hash chains in SAODV*

With regard to the hash chain, a hash function h is known by nodes participating in the routing process at the outset. This function is chosen by the source node. The selected hash function is considered as a non-mutable field and is protected by a signature. Note, in this case, that the obfuscation principle for function h does not consider Kerckhoffs' principle [PET 11]. However, this principle remains acceptable in a real-time context; if a hash function based on a function composition is used (as described in [NDO 11]), then a hash function known in advance by all acceptable nodes is still permissible. The time needed for an attacker to identify the function may be sufficiently long to prevent the malicious element from damaging the integrity of routing messages (which are executed in real time).

The construction of the hash chain is illustrated in the example below, using Figure 3.7.

In this example, the source node S is seeking a route to a destination D. It thus carries out the following operations:

1) Choice of the function $Hash\_Function = h$.

2) Choice of a random number for the $seed$.

3) Choice of the *Max_Hop_Count* field, which depends on the diameter of the network (and may be equal to TTL = 64).

4) Apply Hash = seed.

5) Calculate $Tophash = h^{Max\_Hop\_Count}(seed)$. In this expression,

- h is the hash function

- $h^i(data)$ is the result of the application of function h, i times the $data$

On reception of the message by node N1, the following operations are carried out:

1) Calculation of $h^{Max\_Hop\_Count-hopcount}(Hash)$ and verification of the equality with the value in the $Tophash$ field. As $Hash$ is equal to $seed$ and the current hop count is 0, we obtain $h^{Max\_Hop\_Count-0}(seed) = Tophash$. This equality indicates that the hop count has not been modified.

2) If verification is successful, node N1 increases the hop count by 1 and applies the hashing function to the previous value of $Hash$ (equal to $seed$) before transmitting it to node N2.

The new value of Hash is thus $Hash = h(seed)$.

The following operations are carried out when the message is received by node N2:

1) Calculation of $h^{Max\_Hop\_Count-hopcount}(Hash)$ and verification of the equality with the value of the $Tophash$ field. As $Hash$ is equal to $h(seed)$ and the current hop count is equal to 1, we have $h^{Max\_Hop\_Count-1}(h(seed)) = Tophash$. This equality indicates that the hop count has not been modified.

2) If verification is successful, node N2 increases the hop count by 1 and applies the hashing function to the previous value of $Hash$ (equal to $h(seed)$) before retransmitting the packet.

The new value of Hash is thus $Hash = h^2(seed)$.

It is important to note that the decision to differentiate between mutable and non-mutable fields is justified by network performance requirements. If all fields are signed, this may generate a significant overhead, leading to time delays. For this reason, SAODV does not sign all messages at each retransmission, but simply verifies the integrity of the only variable field (the hop count) using a hashing function (which is less costly).

3.6.4. *Wormhole attacks*

The wormhole attack is a serious attack that targets the routing process. It is effective against reactive routing protocols using hop counts as their route selection metric. Note that a wormhole attack may be carried out using two or more attackers to transport packets from one location to another within a network. The attack creates a disturbance in the packet routing process, as distant nodes are led to believe that they are neighbors. Genuine nodes are obliged to follow the routing protocol algorithm, choosing the shortest available route, and thus transmit packets through the wormhole.

Figure 3.9 shows an example of a wormhole attack. In this case, nodes Dr1 and Dr4 are led to believe that they are neighbors. Attacker A1 transfers all packets from node Dr1 directly to Dr2, via a second attacker A2. Node Dr4 thus believes that Dr1 is in range, that is, a neighbor. All control packets (Hello, RREQ, RREP, RERR) are then transmitted along this route. Even with no knowledge of cryptographic keys or of the selected hash function, attackers are able to damage the integrity of the network by transferring control packets, and subsequently capture data traffic.

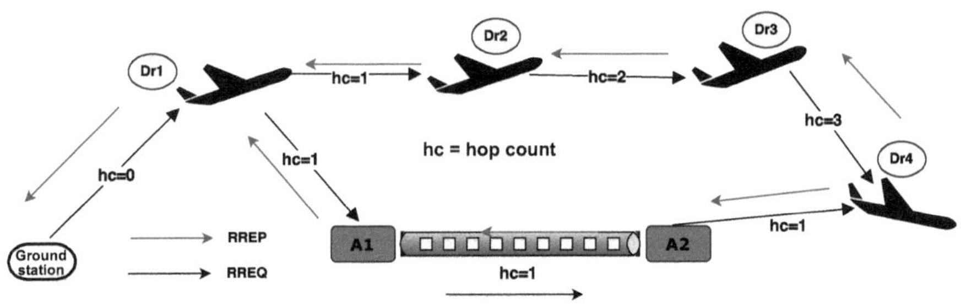

Figure 3.9. *Illustration of a wormhole attack. For a color version of this figure, see www.iste.co.uk/maxa/software.zip*

3.6.5. *Single attacker variant*

The SAODV protocol is also vulnerable to single attackers which operate by capturing, but not modifying, routing packets [ZAP 02b]. This attack is possible due to the fragility of the hash function approach used in SAODV. In Figure 3.10, an attacking node A, not in possession of the network hash function, does not increase the hop count in order to reduce the number of hops on a route passing through this node. To do this, node A places itself within range of two targets, Dr1 and Dr3. The distance between Dr1 and A and between A and Dr3 cannot exceed the maximum range covered by a single hop. This maximum value is noted D_{max}.

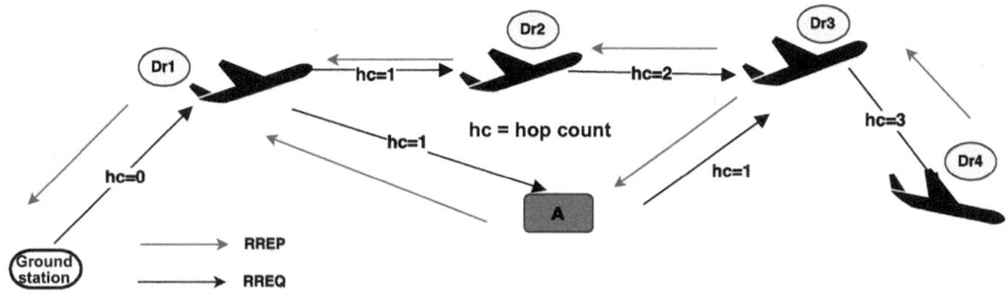

Figure 3.10. *Simplified wormhole attack by a single attacker. For a color version of this figure, see www.iste.co.uk/maxa/software.zip*

Applying the hash chain algorithm used in SAODV, the nodes exchange the following messages:

Dr1 $\rightarrow$ Dr2: (64, Top_hash, Hash_function, Max_hop_count, signature, Hash) where:

– 64 indicates that the message is a request packet.

– $Hash\_function = h$

– $Tophash = h^{Max\_hop\_count}(seed)$

– $Hash = h(seed)$

Following verification, node Dr2 sends the following message:

Dr2 $\rightarrow$ Dr3: (64, Top_hash, Hash_function, Max_hop_count, signature, Hash), where:

– $Hash\_function = h$

– $Top\_hash = h^{Max\_hop\_count}(seed)$

– $Hash = h(Hash) = h(h(seed))$

If an attacker is present, node Dr3 will receive two types of messages:

1) The first message, sent by Dr2, will be authenticated by comparing Top_hash with the value $h^{Max\_hop\_count-2}(h^2(seed)) = Top\_hash$

2) The second message, sent by A1, will be authenticated by comparing Top_hash with the value $h^{Max\_hop\_count-1}h(seed) = Tophash$

Consequently, the presence of an attacker on the route passing through node A will not be detected, and this route will be preferred as the hop count is lower.

3.6.6. *State of the art: solutions for defense against wormhole attacks*

Several approaches [PAT 15] have been proposed to defend against wormhole attacks in MANET networks. These different approaches are summarized in Table 3.6.

Methodology	**Protocol**	**Detection of wormhole attacks**	**Prevention of wormhole attacks**
Wormhole detection algorithm based on AODV [YAO 13]	AODV	✓	✗
Directional antenna approach [HU 04a]	AODV	✓	✗
ConSetLoc [NIU 12]	Protocol using the hop count metric	✓	✓
Packet leashes [PER 03c]	Protocol using the hop count metric	✓	✓

Table 3.6. *Summary of selected existing options for defense against wormhole attacks*

The authors of [PER 03c] propose a mechanism based on the packet leash principle. This is based on the addition of supplementary information (of a temporal or geographical nature) to a packet in order to detect the presence of wormholes, with the aim of limiting the maximum travel distance for a packet. For this to be possible, the nodes in a network must be synchronized. The general principle of packet leashes is based on a calculation of the distance traveled by a packet, using either node positions, in the case of geographical leashes, or the time difference between nodes in the case of temporal leashes. As we shall see, these mechanisms offer no defense against wormholes, essentially due to the introduction of a non-deterministic and non-negligible processing delay.

The temporal leash mechanism uses transmission times to detect wormhole attacks. It carries out a wormhole search during the route discovery process, calculating the transmission time between two successive nodes along the whole of the route. Expression [3.1] indicates the calculation method used by nodes:

$$te = ts + L/C - \delta \qquad [3.1]$$

– te: expiry time

– ts: packet sending time

– C: propagation speed of a wireless signal, more or less equivalent to the speed of light

– L: maximum distance a packet can travel for transmission between neighbors (one hop)

– δ: maximum clock difference between two nodes.

Detection is based on the fact that the transmission time between two nodes connected by a wormhole is considerably higher than that between two genuine nodes. The drawback to this approach is the high probability of generating false positives, as the difference is essentially dependent on the following temporal variables:

– The transmission time, which is the time taken to upload the packet for sending to the wireless network. This value depends notably on the type of connector between the system responsible for the payload and the antenna. It also depends on the size of the packets and on the size of the queue in case of interference. The system between the onboard calculator and the antenna, including the modem, also adds a delay which may be greater than the propagation time.

– The processing time in a node corresponds to the time taken to process a message, and depends on the node's processing capacity (CPU).

In order for the temporal leash algorithm to work, the synchronization method must therefore take account of non-deterministic variables which are dependent on the material characteristics of the hardware in question.

The geographical leash mechanism consists of adding the position of an emitting node into the request, allowing the node which receives the packet to measure the distance between it and the previous node (see expression 3.2). Assuming that the nodes are synchronized, each node is able to detect cases where the packet has traveled further than expected. The basic formula used is as follows: *let S be the source node, R the receiving node, ts and tr the sending and receiving times respectively, and V the upper bound of the node speed. The distance between S and R must respect the condition:*

$$dsr \leq (ps - pr) + 2V(tr - ts + \alpha) + \delta \qquad [3.2]$$

This expression indicates the distance between S and R. However, the approach is of limited use in cases where the destination node R is outside of the zone covered by node S due to the presence of an obstacle between the two. In this case, the packet must pass through an intermediary node, N1, as illustrated in Figure 3.11.

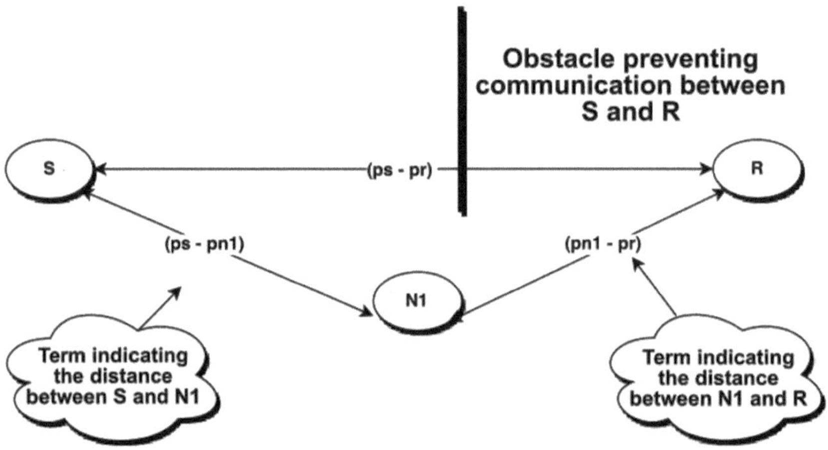

Figure 3.11. *Limitations of use of the geographical leash mechanism*

In this figure, node S sends a packet, inserting its position (noted ps). In the absence of obstacles, the distance between S and R may be verified by the condition described in formula [3.2]. Otherwise, if the packet is first transmitted to N1, then retransmitted to R, this condition is no longer respected, even in the absence of wormholes. The limitations of this mechanism are demonstrated below by comparing the distance between nodes S and R, with and without an obstacle.

In cases with an obstacle, we have:

$$(ps-pr)^2 = (ps-pn1)^2+(pn1-pr)^2-2(ps-pn1)(pn1-pr)\cos(\alpha) \quad [3.3]$$

α indicates the angle between the segments connecting nodes S and R and S and N1.

Since:

$$(ps-pn1)^2+(pn1-pr)^2 > (ps-pn1)^2+(pn1-pr)^2 - 2(ps-pn1)(pn1-pr)\cos(\alpha)$$

then:

$$(ps-pn1)^2+(pn1-pr)^2 > (ps-pr)^2$$

Hence,

$$((ps-pn1)+(pn1-pr))^2 > (ps-pn1)^2+(pn1-pr)^2 > (ps-pr)^2$$

which implies that:

$$(ps-pn1)+(pn1-pr) > (ps-pr) \quad [3.4]$$

Dividing the terms of this equation by the propagation speed and taking β and σ to be two non-zero variables, we have:

$$\left\{\begin{matrix} tr-ts=\beta \rightarrow with \quad obstacle \\ tr-ts=\sigma \rightarrow without \quad obstacle \end{matrix}\right\}$$

This implies that $\rightarrow \beta > \sigma$

Considering this expression in [3.5], we obtain:

$$(ps - pn1) + (pn1 - pr) + \beta > (ps - pr) + \sigma \qquad [3.5]$$

Applying [3.5] to [3.2]:

$$(ps - pn1) + (pn1 - pr) + 2V((tr - ts)_{obstacle} + \alpha) + \delta > ||ps - pr|| + 2V((tr - ts)_{without\_obstacle} + \alpha) + \delta$$

Thus:

$$dsr_{obstacle} > dsr \qquad [3.6]$$

which proves that the mathematical representation [3.2] does not work in the presence of an obstacle.

Another proposal is made in [YAO 13], in the form of a modification to the AODV protocol designed to identify several different routes between a source and a destination. The authors consider that a route including one or more wormhole nodes will be characterized by a significantly lower hop count than the other routes. However, they show that in networks with low node density and where multiple routes cannot be found, this approach does not offer protection against wormhole attacks.

The authors of [HU 04a] propose a method based on the use of a directional antenna. As these antennas turn toward the direction of origin of a packet, they can help to verify the identity of a node at a distance of two or more hops. Using this mechanism, during the neighborhood discovery phase, each node indicates the direction from which messages have been received. Each node then only accepts packets received in the opposite direction. Evaluations have shown this type of mechanism to be effective; however, it requires the use of specific hardware, that is, a directional antenna.

[NIU 12] features an approach known as *ConSetLoc*, based on an evaluation of the relationship between the hop count and the geographical distance between nodes. This approach has been used for fixed probe networks, in which nodes do not change position during execution of the routing algorithm. In this case, a static interval may be used to represent the relative distance between nodes. Performance evaluations have shown the advantages of this approach over other existing methods. However, the results obtained depend on the static position of nodes. It would be interesting to consider ways to adapt this approach for use in a more dynamic situation such as a UAANET network.

3.6.6.1. *Summary of existing work*

The packet leash mechanism may be used to detect wormhole attacks if the connectivity between two nodes is not interrupted by an obstacle, obliging them to pass through an intermediary node. The ConSetLoc approach provides the same service, but the protocol has yet to be tested in a dynamic environment in which node positions are modified in real time. For this reason, we opted for a solution which combines the two approaches, based on an improved form of the geographical leash algorithm. In this way, we have attempted to create a correspondence between the relative distance between two nodes and the hop count included in the packet. Using our approach, the travel distance for each message is limited; each message has a time stamp and gives the location of its source. The destination node calculates the distance relative to its own location and clock. This relative distance is then compared with the hop count contained in the packet. If the distance traveled by the packet corresponds to the expected hop count value, this indicates that there are no wormholes in the network.

Note that this approach can be used in other types of MANET network, but it is particularly interesting in the case of UAANET networks as the nodes can be synchronized.

3.6.7. *A new method for detecting and defending against wormhole attacks*

The first approach used to detect attacks of this type is based on the geographical leash principle. As we have seen, the initial form of this approach, as presented in [PER 03c], creates false positives and is not always effective in detecting wormhole attacks (notably if an obstacle is present between two nodes that are usually neighbors). We have thus formalized another approach to create a distance interval. This consists of evaluating the relationship between the relative geographical distance between two neighboring nodes and the hop count contained in the packet. Our mechanism is based on a non-deterministic distance, bounded by a maximum value D_{max}, representing the maximum range of an omnidirectional node antenna.

To do this, each node includes its position when sending a control packet. On receiving a packet, a node calculates the relative distance between it and the sender, using the formula for Euclidean distance in a three-dimensional environment (see equation [3.9]).

In order for this algorithm to work, the nodes need to be synchronized. In a UAANET network, each node has a GPS that allows it to know and to communicate its position to the GCS in real time. This enables synchronization. Furthermore, the DT18 drones used in this case possess a high-capacity 8-core processor, offering an added advantage in terms of position calculations.

Below, we shall analyze the wormhole attack and show the correspondence between geographical distance and hop counts used to detect these attacks. The notation used in Figure 3.12 is given in Table 3.7.

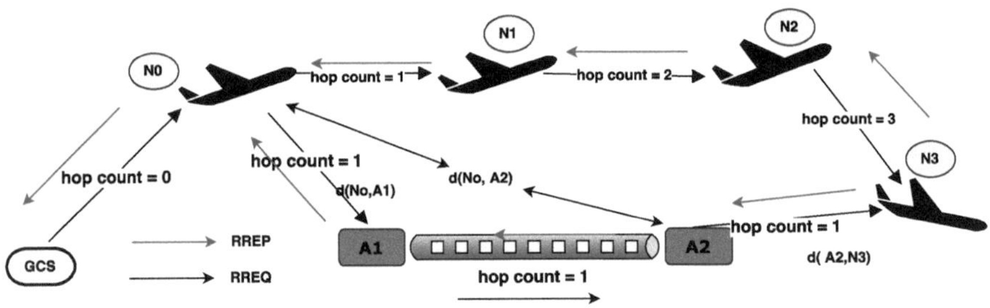

Figure 3.12. *Wormhole-type attack with two attackers. For a color version of this figure, see www.iste.co.uk/maxa/software.zip*

We may state that:

– the wormhole attack creates a tunnel, which transmits packets over a route with a lower hop count than other connections.

– D_{max} represents the maximum radio range of a node. These data can be obtained at radio level, as it is possible to adjust the emitting and receiving signal powers of the wireless adapter.

We have: $\forall i, j \in [0, n], R_{ij} \leq D_{max}$ (*with n the maximum number of nodes in the network.*)

Parameter	Description
hc	Hop counter
D_{max}	Maximum hop distance
R_{ij}	Real distance between two nodes i and j
$d(No, A1)$	Distance between N0 and A1
$d(A1, A2)$	Distance between the two attackers
$d(A2, D)$	Distance between the second target and the second attacker
$d(No, A2)$	Distance between N0 and A2
Pi	Sum of the distance traveled by a packet A at an instant t
T	Sum of the total distance traveled over the genuine route
Dw	Total length of the route passing through the wormhole

Table 3.7. *Notation table*

Thus, the connectivity between two genuine neighboring nodes, noted c, is represented by the following conditions:

$$c(i,j) = \begin{Bmatrix} 1\ if\ R_{i,j} \leq D_{max} \\ 0\ if\ R_{i,j} > D_{max} \end{Bmatrix} \qquad [3.7]$$

In the presence of a wormhole, this condition is not respected, and two nodes separated by a distance greater than D_{max} may consider themselves to be neighbors.

$$c(i,j) = \begin{Bmatrix} 1\ if\ R_{i,j} \leq D_{max} \\ 1\ if\ R_{i,j} > D_{max} \end{Bmatrix}$$

Next, for the tunnel to be useful to an attacker, the second attacker A2 should not be within the zone covered by N0. If A2 is within the range of N0, the tunnel is pointless, given that its purpose is to create a connection between two nodes separated by a distance greater than D_{max}. Hence:

$$d(No, A1) \leq D_{max}$$

$$d(A2, N3) \leq D_{max}$$

$$d(No, A2) > D_{max}$$

$$d(A1, N3) > D_{max}$$

We thus obtain:

$$d(No, A2)^2 = d(No, A1)^2 + d(A1, A2)^2 - 2d(No, A1)d(A1, A2)cos(\gamma)$$

where γ is the angle formed by the segment traced by (N0, A1) and (A1, A2).

Hence,

$$d(No, A1)^2 + d(A1, A2)^2 > Dmax^2 \rightarrow d(A1, A2)^2$$
$$> Dmax^2 - d(No, A1)^2$$
$$d(A1, A2)^2 > (D_{max} - d(No, A1))(D_{max} + d(No, A1))$$
$$d(A1, A2)^2 > (D_{max} - d(No, A1))(D_{max} - d(No, A1))$$

Hence,

$$d(A1, A2) > D_{max} - d(No, A1)$$

Given that:

$$Dw = d(A1, A2) + d(No, A1) + d(A2, N3)$$

We obtain:

$$Dw > D_{max} \quad [3.8]$$

We can use this distance inequality to compare the hop count in the packet with the expected hop count, based on the distance between the two neighboring nodes. As node positions are included in packets, each drone has the ability to calculate the distance traveled by each packet. The distance between two nodes i and j can be found using the following formula:

$$Rij = \sqrt{(xj - xi)^2 + (yj - yi)^2 + (zj - zi)^2} \quad [3.9]$$

where (x, y, z) represent the set (latitude, longitude, altitude). For example, xj represents the latitude of node j, and xi represents the latitude of node i.

In the case of a wormhole attack, there will be an anomaly between the distance and the hop count.

Given that

$$T = \sum_{i=0,j=0}^{n} R_{i,j}$$

– when node N0 sends a packet, we have $T0 = R_{01}$, corresponding to hc = 1;

- R_{01} indicates the distance between N0 and N1;

– when node N1 sends the packet, we have $T1 = T0 + R12$, corresponding to hc = 2;

- R_{12} indicates the distance between N1 and N2;

– when node N2 sends the packet, we have $T2 = T1 + R23$, corresponding to hc = 3;

- R_{23} indicates the distance between N2 and N3.

Table 3.8 is then created in order to detect wormhole attacks.

Value of T	**Hop count hc**
$0 < To \leq Dmax$	0
$D_{max} < T1 \leq 2Dmax$	1
.....	.....
$(n-1)D_{max} < T_{n-1} \leq (n+1)Dmax$	n-1

Table 3.8. *Correspondence table between geographical distance T and hop count*

From this table, we obtain:

$$\frac{T}{Dmax} - 1 \leq hc < \frac{T}{Dmax} + 1 \quad [3.10]$$

To detect a wormhole attack, we can either use the correspondence table or use this inequality to see whether the hop count falls within the interval given above.

EXAMPLE.– Let us consider Figure 3.12 as an example of the way in which this mechanism operates.

1) In the case where node N3 receives a message from N2:

- N3 calculates the distance R_{23} using formula 3.9;

- It then calculates the current value of the sum of distances T as follows: $T2 = R_{01} + R_{12} + R_{23}$

- Hence, equation 3.10 becomes:

$$\frac{R_{01}+R_{12}+R_{23}}{Dmax} \leq hc < \frac{R_{01}+R_{12}+R_{23}}{Dmax}+1$$

- Then, since $T2 = R_{01}+R_{12}+R_{23}$, we have:

$$\frac{T2}{Dmax}-1 \leq hc-1 < \frac{T2}{Dmax} < \frac{T2}{Dmax}+1$$

- Based on Table 3.8, we should therefore have $hc = 2$ as $2D_{max} < T2 \leq 3D_{max}$

- Comparing this result with the current value of *hc*, which is equal to 2 at this point, we see that the packet has not traveled over a wormhole link.

2) In the case where this message comes from A2:

- N3 calculates the distance traveled by the packet, which is equal to $T2 = Dw$;

- Equation 3.10 thus becomes:

$$\frac{Dw}{Dmax}-1 \leq hc < \frac{Dw}{Dmax}+1$$

- Given that $Dw > D_{max}$, we have: $Dw = \gamma D_{max}$ with $\gamma > 1$. Hence: $0 < \gamma - 1 \leq hc < \gamma + 1 < 2\gamma \Rightarrow 0 < hc < \gamma + 1$

- Comparing this result with the hop count value $hc = 0$ found in the packet (which has not been modified by the attacker), we see that the packet has been transmitted by one or more wormhole attackers. Generally speaking, as long as $Dw > D_{max}$ (which is always the case if a wormhole is present), the attack will be detected.

This example shows how our proposed mechanism may be used to detect and defend against wormhole attacks.

3.6.8. *Defense mechanism for single-attacker wormhole attacks*

The mechanism presented above may also be used as a defense against the attack described in section 3.6.5. As we explained earlier, this attack consists of transferring a packet obtained through illicit listening to a neighboring node. In Figure 3.10, the node transfers the unmodified packet toward N3.

An additional element is used in the mechanism to defend against this attack. Once we know that a neighbor is within a distance of less than D_{max}, we can trust this node, including its identity in a hashing function. The principle used for identification here is to include the identity of the next node[5] in the hashing function for each node. The following operations are carried out, using Table 3.9.

Field	Value
Type	64
Hash function	H
Signature	Signature of non-mutable fields
Hashnew	A value of fixed size obtained using the operation H[Current_node, Next_node, Previous_hash chain]. The first two variables indicate the IP address of the source node and of the next node. The third variable is the previous hash value
Hashold	Previous hash value

Table 3.9. *Different fields in a request packet*

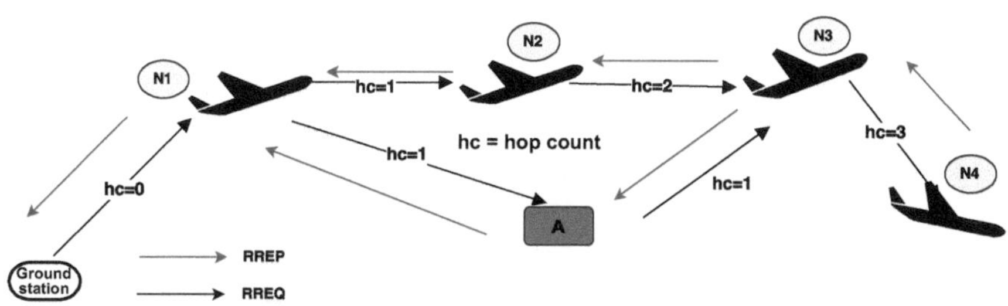

Figure 3.13. *Simplified single-attacker variant of the wormhole attack. For a color version of this figure, see www.iste.co.uk/maxa/software.zip*

5 The identity may be an IP address or a public key

Node N1 carries out the following operations:

– selection of a function H to use;

– calculation of $Hashold = H(seed)$ using a random value chosen by the emitting node as the seed;

– calculation of $Hashnew = H(N1, N2, Hashold)$, with N1 the address of the following node;

– send message N1 $\Rightarrow N2 : [64, H, signature, Hashnew, Hashold]$.

On receiving the packet, node N2 carries out the following operations:

– integrity verification, calculating $Hashverifier = H[Previous\_node, Current\_node, Hashold]$, and verification of correspondence with $Hashnew$. Based on the one-way property of the hash function, any modification of the message would create a difference in Hashverifier;

– $Hashverifier = H[N1, N2, Hashold] = Hashnew$, showing that the route has not been retransmitted by an attacker;

– calculation of the new value of $Hashold = Hashnew$;

– calculation of $Hashnew = H[N2, N3, Hashold]$;

– send message N2 $\Rightarrow N3 : [64, H, signature, Hashnew, Hashold]$.

As attackers simply transfer packets without modifying them in this case, attacks will be detected:

1) For a genuine route, we have $Hashverifier = H[N2, N3, Hashold]$, which is equal to Hashnew.

2) For a wormhole route, we have $Hashverifier = H[N1, N3, Hashold] \neq Hashnew$, so node N3 will detect an anomaly in the network.

The operation is repeated until the packet reaches the intended recipient, and the same operation is carried for the RREP packet traveling back to the source. The exact value of the hop count can be deduced from the number of times the hash is used for verification. It may also be included in calculating the hashing chain.

Note that this mechanism may also be used to detect wormhole attacks by two attackers. Detection is possible in any case where the wormhole considers two non-neighboring nodes to be neighbors. However, it cannot detect the presence of wormhole attacks independently, as it cannot tell if a neighboring node is a wormhole node. The first approach, described in section 3.6.7, is able to do this.

3.6.9. *Limitations of the SUAP protocol*

The mechanisms used for message authentication in SUAP are based on the SAODV protocol. The use of a digital signature for non-mutable fields forms an efficient defense against any type of modification attack and against the entry of fake packets into the network. For non-mutable fields, the hashing chain principle can also be used to ensure integrity of the hop count for each retransmission. However, this principle is known to be vulnerable to wormhole attacks, and for this reason, we proposed a mechanism based on the use of the hash function in SAODV.

However, the hash function principle falls down in failing to consider Kerchoffs's principle. In SAODV, a hash function table is created, and is only known by genuine nodes, allowing them to identify the function chosen by the source node. This may be seen as a form of security by obfuscation, as the function is only initially known by genuine nodes.

If an attacker is able to gain knowledge of this function, a number of attacks on the SUAP protocol are possible.

– An attacker may use the hash function to increase the hop count in order to prevent a packet from being transmitted over an optimum route. For example, an attacker may forcibly increase the hop count for a genuine route so that packets are transmitted over a different, non-optimal route, thus reducing network performance.

– One or more wormhole attackers may also modify the value of the Hashnew and Hashold stamps to avoid detection. This more sophisticated form of attack consists of not only creating a tunnel, but also modifying the packet exchanged through the tunnel. Attacks of this kind have not been considered for this study, as we have only considered wormholes which do not modify packets, as defined in [YAO 13]. In the case of a modification attack, if the attacker manages to deduce the value of Hashnew from Hashold by taking account of the identity of genuine nodes, packets passing through a wormhole may not be detected.

However, this type of attack involving value modification has limited effects in a real-time context, particularly when a more robust hash function is used. If the hash function in question results from a composition of functions (as described in [NDO 11]), the function may be public, as the time needed for an attacker to calculate the function would be too high for them to have an impact on the integrity of routing messages (executed in real time during the mission).

This specific case shows that the SUAP protocol is sufficient to guarantee detection of wormhole attacks in a UAANET context. However, in cases where the SUAP protocol is used for other families of MANET networks (such as sensor networks), it may not be sufficient to provide wormhole protection. In this case, a different approach to detection may be useful. In a low-density MANET network, for example, we may consider using a mechanism based on the relationship between the hop count and relative distance, as described in section 3.6.2.2, for all routing packets[6].

3.7. Using the AVISPA tool to verify the security properties of the SUAP protocol

Formal verification of the SUAP protocol using AVISPA is a two-part process.

1) Firstly, we need to specify the protocol using a high-level specification language, expressing the message exchanges taking place by different agents in the protocol in mathematical terms. This is done using HLPSL. This language enables us to create an abstraction of the protocol and to create a detailed specification of its different security aspects (such as the roles of nodes, cryptographic operators, intruder models, etc.). This specification is then automatically translated into an intermediary format, IF, using a translator (HLPSL2IF).

6 In cases with a high node density, calculating the relation between the hop count and relative distance may not be practical, as the time and overhead associated with position signature mechanisms for each retransmission are non-negligible.

2) The second stage is to verify the specification of the SUAP protocol with different backends. AVISPA uses four different modules (backends) using the IF as input, allowing four different analyses of the same protocol. These backends, integrated into AVISPA, make it possible to verify security properties as required.

It is also important to consider intruders in the context of the verification process, that is, to specify their behavior. The most common hypotheses are based on the Dolev-Yao model [CER 01]. This model uses a hypothesis of perfect encryption and considers the intruder to be within the network. Perfect encryption reduces the intruder's decryption capacities; we thus consider that an intruder will be unable to obtain information without the decryption key. The second hypothesis relates to the intruder's abilities. Specifying that the intruder is within the network means that agents send their messages to the intruder. This intruder may then choose whether or not to transmit these messages to their intended recipient. The intruder may also send a message formed from a combination of its knowledge to any node in the network, encrypted using a key which it has been able to obtain.

Below, we shall consider the specification we developed for a formal security study of the SUAP protocol.

3.7.1. *Application of the SUAP protocol*

To verify the SUAP protocol, we created a specification for its possible behavior. This specification is intended to verify whether the protocol is able to guarantee message authentication in the desired manner. However, we were unable to create a specification for mechanisms defending against wormhole attacks, as AVISPA does not include the integer comparison capacities needed to take account of the relation between hop count and relative distance. This property was therefore verified during the protocol validation study, described in detail in the following section.

It is helpful to describe a number of characterization tools used in specifying the SUAP protocol.

Communication node behaviors have been grouped into roles. All nodes in the UAANET network carrying out the same actions share the same role. In our case, there are three distinct roles: the source node, neighboring nodes which participate in packet transfer, and the destination node. Each agent (the term used in AVISPA to denote a node) is associated with a role. Within a role, we describe an initial state, a state change and the information to modify with each state change. The source node specification is shown in Figure 3.15.

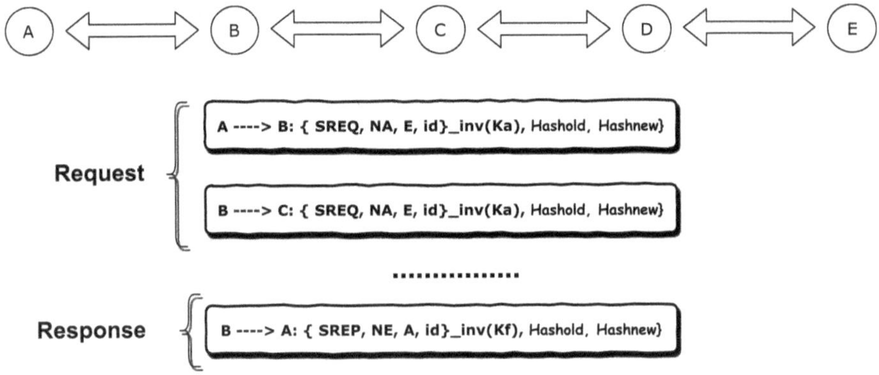

Figure 3.14. *Example of an exchange in the SUAP protocol involving five nodes*

```
%****************************************Source****************************************************%
role sourceNode(
                              A,B,C,D,E: agent,
                              Ka, Kb, Kc, Kd, Ke : public_key,
                              Hf : function, %Hash function
                              KeySet : (public_key) set,
                              RCV, SND : channel(dy)
                              )
        played_by A def=
                local
                        State : nat,
                        REQstatic, REPstatic : text,
                        SREQ, SREP : protocol_id,
                        Hashnew, Hashold, Na, Nd : message
                init
                State := 0

                transition
                        step1.
                              State = 0  /\ RCV(start) =|>
                              State':= 8 /\ Na' := new()
                                          /\ Hashnew' := new()
                                          /\ Hashnew' := Hf(A,B,Hashold)
                                          /\ SND(SREQ.REQstatic.Na'.Hashold.Hashnew'.{Na'.SREQ.REQstatic}_inv(Ka))
                                          /\ witness(A,B,Na,Na')

                        step2.
                              State = 8 /\ RCV(SREP.REPstatic.Nd'.Hashold.Hashnew'.{Nd'.SREP.REPstatic'}_inv(Kb).H) /\ in(Kb, KeySet)/\
Hashnew'=Hf(A,B,Hashold)=|>
                              State':=9 /\ wrequest(A,B,Na,Na')

end role
%****************************************R1:intermediate_node1****************************************************%GGG
```

Figure 3.15. *HLPLS specification of the source node. For a color version of this figure, see www.iste.co.uk/maxa/software.zip*

The properties of the protocol to analyze via these representations are:

– authentication of the NA field, Tophash and the identifier of node E during the discovery phase. Specifically, node E must be able to verify the authenticity of the couple (NA, E);

– the integrity of the hashnew field. Each node receiving the packet calculates the Hashverifier variable, as explained above, and compares it with the current value of hashnew;

– authentication of the ND field and the identifier of node A during the discovery phase (response packet).

We thus verify both the end-to-end authentication of static fields and the integrity of mutable fields. Our HLPLS specification allows us to do this by telling node A to initialize the session and send messages containing packet identifiers and hash values. This message is then transmitted to nodes B, C and D, until a node with knowledge of the recipient responds with an RREP packet. When an intermediary node receives a packet, it verifies the certificate and the signature of the previous node. When the packet reaches its destination, the variable couple (NA, E) is verified. If verification is successful, the destination node generates a response packet, including the packet identifiers, the IP address of the source node, the certificate, and the full, signed message. As in the case of the request packet, this packet also includes the variable couple (NE, A), which is verified by the intermediary nodes.

3.7.2. *Analysis of the specification of the SUAP protocol*

The result shown in Table 3.10 indicates that our protocol provides the expected security services, that is, end-to-end authentication and message integrity guarantees (hop-by-hop authentication). However, considering the representation below, we see that an intruder situated within the range of a genuine node would be able to obtain and retransmit messages without modification. This type of attack was discussed in section 3.6.5. The proposals made in section 3.6.7 offer a means of defense against this attack, as we shall see in the following section. We will also present the results of evaluations carried out using simulation and in real experiments with regard to this attack type.

1) $A \rightarrow I : \{SREQ, E, Na'\}\, inv(Ka), Hash, certA$

2) $A \rightarrow B : \{SREQ, E, Na'\}\, inv(Ka), Hash, certA$

3) $I \rightarrow B : \{SREQ, E, Na'\}\, inv(Ka), Hash, certA$

4) $B \rightarrow I : \{SREQ, E, Na'\}\, inv(Ka), Hash, certA, certB$

OFMC output	ATse output	SATMC output
%% Number of warnings: 9 % OFMC SUMMARY **SAFE** DETAILS BOUNDED_NUMBER PROTOCOL /home/span/old.if GOAL as_specified BACKEND OFMC COMMENTS STATISTICS parseTime: 0.00s searchTime: 0.04s visitedNodes: 5 nodes depth: 4 plies	%% Number of warnings: 9 %ATSE SUMMARY **SAFE** DETAILS BOUNDED_NUMBER PROTOCOL /home/span/old.if GOAL As Specified BACKEND CL-AtSe STATISTICS Analysed: 3 states Reachable: 1 states	%% Number of warnings: 9 SUMMARY **SAFE** DETAILS STRONGLY_TYPED_MODEL BOUNDED_NUMBER PROTOCOL /home/span/old.if GOAL %check the HLPSL Specification BACKEND SATMC STATISTICS fixedpointReached 5 steps stepsNumber 5 steps ATTACK TRACE %% no attacks have been found

Table 3.10. *Results obtained using the different AVISPA backends with the SUAP protocol*

In this case, the intruder **I** acts in the same way as a genuine node, transmitting the request toward node B. This request is then transferred by node B toward the next node in the routing table. Node B has no way of knowing that node I is not genuine, as it does not modify the packet.

Whilst it is not possible to find out exactly why this vulnerability is present using AVISPA, we may deduce that it is caused by an illicit listening type of attack carried out by an intruder.

3.8. Implementation of the SUAP protocol

The context in which the SUAP protocol was developed was presented in Chapter 3. We have also analyzed the methodology involved in its implementation. In the previous section, we presented the functions which the protocol is intended to fulfill. Here, we shall consider the technical aspects of SUAP, the implementation choices we made and the techniques used to design the protocol. We shall begin by describing the software architecture of the SUAP protocol and by presenting an overview of its modular breakdown.

3.8.1. *Software architecture of the SUAP algorithm*

The SUAP protocol needs to provide several functions, as described in section 3.6.2.2. First and foremost, it needs to process and route the control packets exchanged between nodes. Secondly, it must ensure that communications can be carried out securely, guaranteeing the authentication and integrity of messages. The provision of these security services should not hinder the operation of the network by poor management of resources. Next, non-functional requirements, such as the safety and operational security of the system, must be respected, following the methodology set out in Chapter 2.

The architecture of the SUAP protocol is shown in Figure 3.16. This illustration formalizes the breakdown of functions into three classes, allowing us to consider each class individually, notably in terms of the sub-set of functions to model. In order to simplify the modeling process, sub-sets linked to other classes have been omitted.

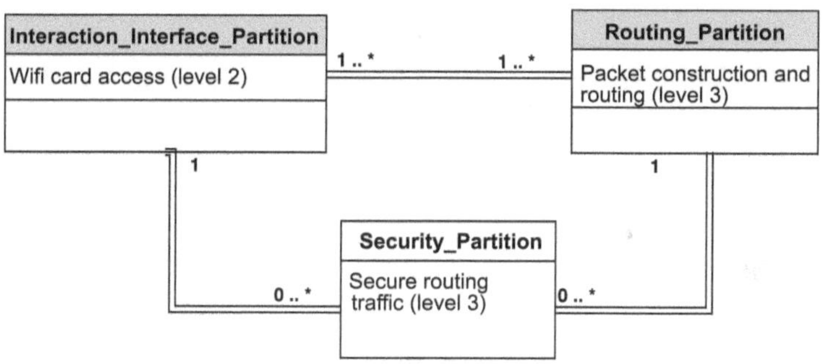

Figure 3.16. *High-level class diagram representing the SUAP protocol. For a color version of this figure, see www.iste.co.uk/maxa/software.zip*

The routing partition class is linked to the protocol's routing functions at network level. Here, we have only considered the IPv4 protocol. This class also manages package filtration on reception, used to select the corresponding process for route discovery and maintenance packets.

The interfacing partition class connects the classes managing routing packets to hardware input and output points. This connection consists of receiving incoming flows from network cards, decomposing the data, and transferring it to the corresponding class. Note that this link is bi-directional, as the interfacing partition also manages the emission of IP packets over the wireless network.

Mechanisms relating to the security of routing packets are implemented by the security partition class.

The tool chain used for our model-driven methodology was presented in the previous chapter. The two main tools used in this case are Simulink® and Stateflow®, which allow us to model and implement the different mechanisms present in SUAP, via the use of block diagrams and state transition diagrams. Each partition class shown in Figure 3.16 is thus associated with a Simulink model, encapsulating these blocks and the associated Stateflow diagrams. The design of this model is described in detail in the following section.

3.8.2. *Modeling the SUAP protocol*

3.8.2.1. *Routing partition*

The first partition class to consider here is the routing class, dedicated to the management of routing packets.

Our conceptualization is based on a generic routing protocol architecture, as shown in Figure 3.17. This generic architecture is made up of several elements: the Demux block, the Maintenance block, the Route Discovery block and the Packet Sending block. The role assigned to each of these blocks will be discussed in detail below in order to clarify their functions.

– Demux block: this block receives packets transmitted by the link layer (for external packets entering the system) or by the transport layer (for packets which have been encapsulated for transport within the network). This block provides internal packet labeling functions using the type field in each packet. These packets are then sent on to the appropriate blocks. For example, on receiving a packet from the transport layer, the Demux block hands it over to the Packet Forwarder sub-block, which consults the routing table to verify the existence of an operational route. If no such route exists, the system hands the packet over to the Route Discovery Block, which generates a request packet and sends it to the Packet_handler block, whence it is sent on to the interfacing partition. This block is implemented by the state machine shown in Figure 3.20. This machine is connected to three network interfaces for receiving packets, and a single output point connected to the next packet processor. In this case, the number of interfaces is arbitrary, and may be increased in future as required.

– The Route Discovery and Route Maintenance blocks determine whether a packet is a request, response, error or Hello message. These four message formats are then assigned to four blocks, each of which is independent but based on either Route Discovery or Route Maintenance. On leaving any of these blocks, the packet is verified to see whether it includes sufficient information to be sent on to the next block, for transmission to the transport or link layer.

– Finally, the packet sending block is used to update the fields of the local packet based on its state. In the case of a packet destined to be processed locally, the destination IP address is verified and the packet is transmitted to the transport layer. If the packet is to be sent over the network, it is transmitted to the link layer.

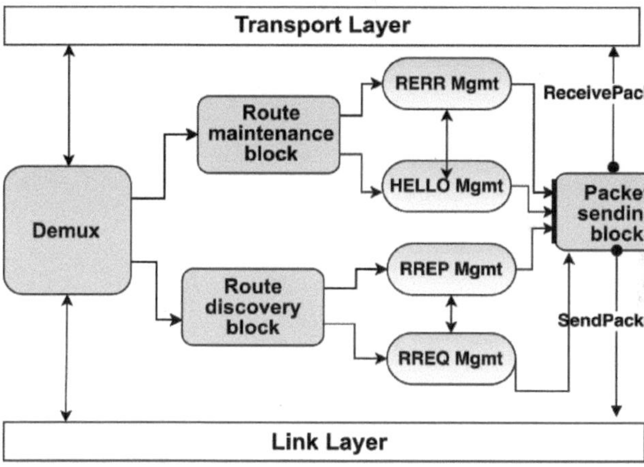

Figure 3.17. *Design of the routing partition class*

The application of this generic architecture is shown in Figure 3.18.

This routing model receives and filters packets through the default port 654. It then verifies the type of the packet, which may be a route discovery or maintenance packet. The packet is transmitted to the corresponding block based on the results of this verification. An example of a Stateflow diagram for request management is shown in Figure 3.19.

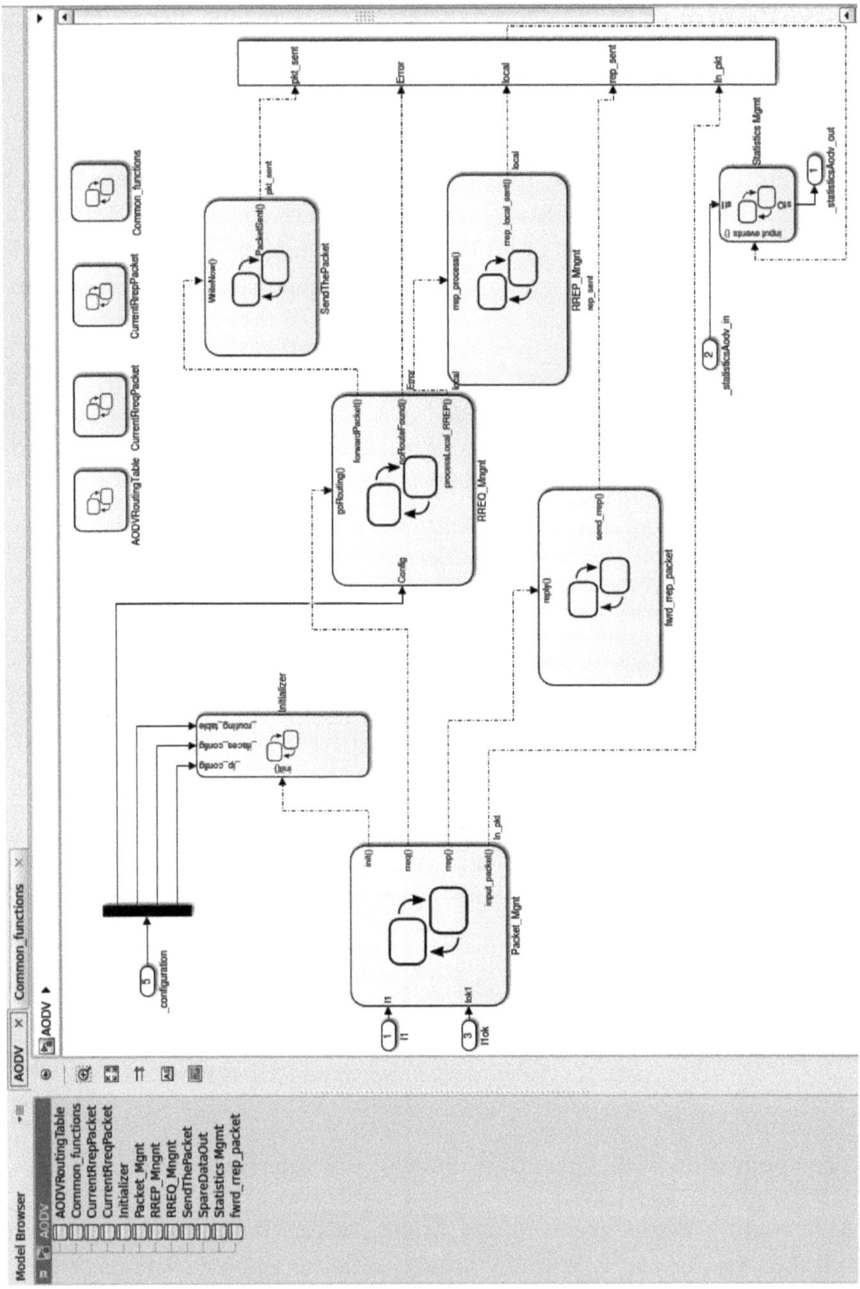

Figure 3.18. *Diagram (simplified) showing Simulink blocks for the routing function*

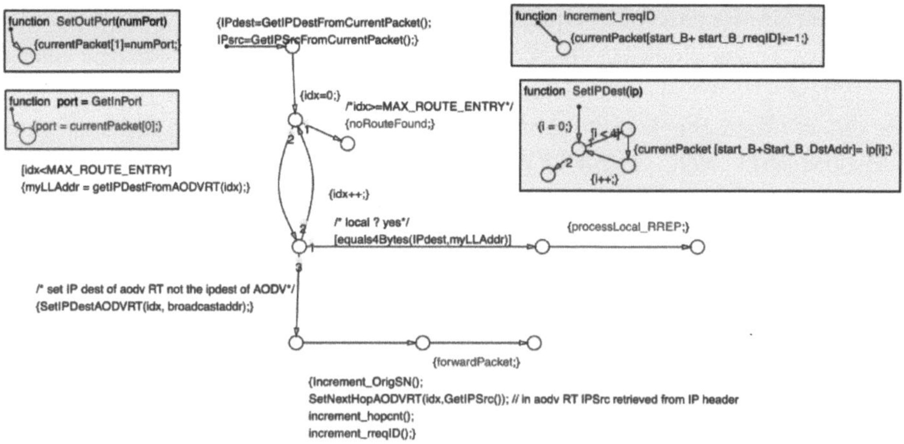

Figure 3.19. *State-transition diagram for packet routing (simplified version). For a color version of this figure, see www.iste.co.uk/maxa/software.zip*

This model contains different logic blocks, each fulfilling specific functions in the packet routing process. The position of these blocks in the diagram is not particularly significant in terms of its translation into source code. The diagram is laid out from left to right for ease of design and reading. The model operates in the following manner:

– First, the system transmits packets to the first block (on the left in Figure 3.18), known as *Packet_Mgmt*. This block sequences incoming packets before transmitting them to the following blocks. It carries out internal labeling of packets in order to transmit them to the corresponding block (route discovery or route maintenance). It is implemented by the state machine shown in Figure 3.20. In this block, packets are retrieved from the global *currentPacket* variable in order to modify their fields. The packet-type field is tested to see whether the packet in question is an RREQ or RREP. There are four possible outputs from this block. The first is the *Init* event, which initializes the configuration of the routing table and network interfaces. This configuration is formalized by an external file, which is transmitted to the program. The second output is *Input_packet*, used to count the number of incoming packets. The program then chooses one of the two blocks (route discovery or maintenance) based on the identified packet type. The two other outputs concern request and response management.

– Next, in the case of a request message, the program process moves on to the *RREQ_Mgmt* block via the *goRouting* event. This block retrieves the source IP address (IPsrc) and the destination IP address (*IPdest*) of the incoming packet. We then check if the number of requests is higher than the maximum threshold *MAX_ROUTE_ENTRY*, set at 16 in our case. If the value is higher than the threshold, the program stops, sending a *NoRouteFound* message as output.

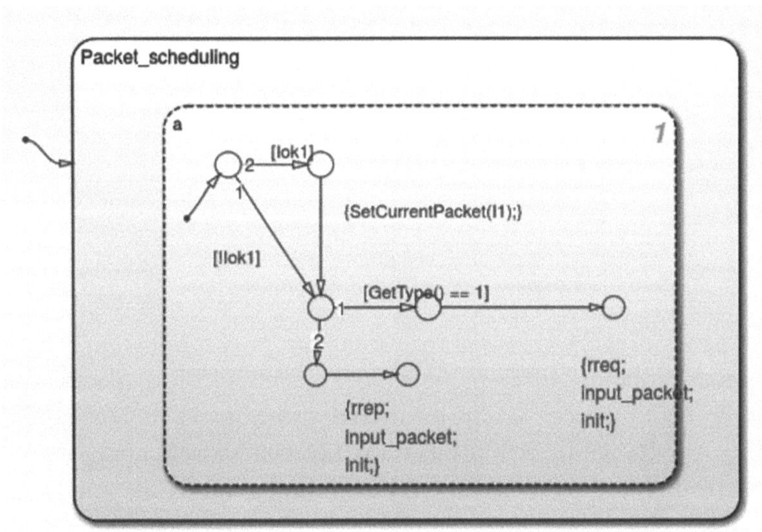

Figure 3.20. *State transition diagram for packet sequencing (simplified version). For a color version of this figure, see www.iste.co.uk/maxa/software.zip*

Next, the IP address of the local node *MyLLAddr* is retrieved and compared with local *IPdest* variables. Depending on the result, the program either hands over to the *processLocalRREP* to send an RREP message back to the source, if a match has been found, or to the *Forwarding* event to broadcast the message to neighboring nodes.

There are three possible processing outcomes from this block:

1) If no route can be found corresponding to the request, the packet is rejected. The *NoRouteFound* signal shown on the diagram is triggered, increasing a variable within the program.

2) If the packet destination address corresponds to that of the node receiving the packet, the routing state machine redirects the packet toward the *ProcessLocal* sub-system. This sub-system is used to respond to requests.

3) If the destination address does not correspond to the local address of the current node, the packet is routed to the output port via the *Forwarding* sub-system, which transmits it to the next node in the routing table.

It is important to note that the design of the structure of the routing table was added at the glue code stage. In this case, we have chosen to use a dynamic table, with input which changes as a function of connectivity. It is thus hard to represent on a Stateflow diagram using static formalized input.

The state transition diagram for request packet routing is shown in Figure 3.19. We see that the algorithm carries out iterative testing of each input element before triggering the associated behavior: *forwaPracket*, *NoRouteFound* or *ProcessLocal_RREP*.

– The *SendThePacket* block allows us to retrieve a packet via the global variable *currentPacket* and to send this packet through the network.

– The *RREP_Mgmt* block creates RREP packets and sends them back to source nodes. Notably, in this block, the *fwrd_rrep_packet* event allows intermediary nodes to update the "TTL", "HopCount" and "IP destination" fields of any RREP packets they receive and to send these packets on to the next node (in the direction of the source node).

– The final block, *Statistics_Mgmt*, receives variables from the other blocks, from which it produces statistics to show whether packets have been received, sent or resulted in errors.

At the end of the process shown in the model, the packet is sent to the routing packet processing block for transmission to the application or layer, marking the end of processing.

3.8.2.2. *Security partition*

A second generic architecture, shown in Figure 3.21, was used to design the security function partition.

This generic architecture is made up of several function blocks, used to identify packets up to the encapsulation of headers used in the secure routing protocol. It is important to note that this model operates in conjunction with the routing model presented above. The routing block model represented by the *Routing verifier* block is added to this partition.

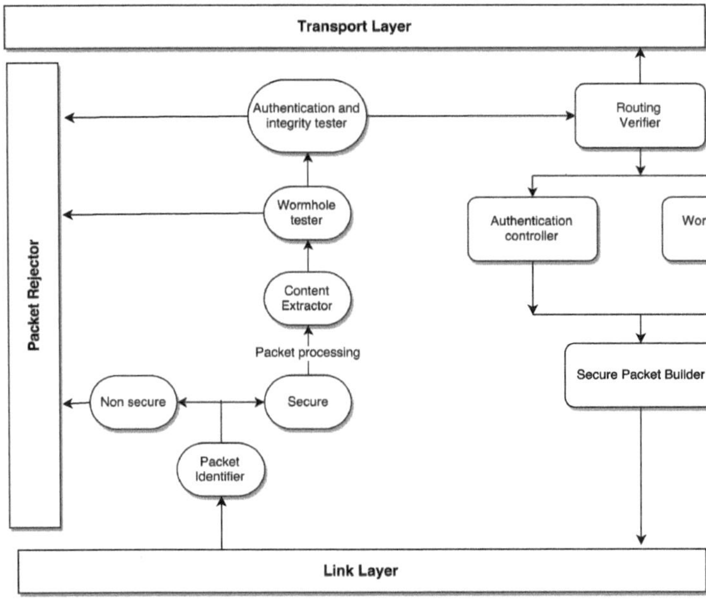

Figure 3.21. *Architecture model used in designing the security partition*

The model operates as follows. On receiving a packet, the *Packet identifier* block eliminates any routing packets without a secure extension. This constitutes a first stage of filtering, designed to eliminate unauthorized packets, and is represented by the *Packet rejector* block.

The next stage is to de-encapsulate the secure packet: this process is carried out by the *content extractor* block. Several different security checks are applied to the packet at this stage. First, the *wormhole tester* block is used to apply distance and hash function calculations, as presented in section 3.6.7. The *wormhole tester* block identifies the packet type (route discovery or maintenance) and applies the appropriate treatment. If a packet fails the *wormhole tester* check, it is rejected via the *Packet rejector* module. Second, the packet is transmitted to the authentication and integrity verification block. The purpose of this block is to verify the signature of non-mutable fields. The hop count is also checked using a hash chain. Once all of these verifications have been carried out, the packet is transmitted to the routing partition, as described above. A simplified illustration of the state transition diagram for the *content extractor* block is shown in Figure 3.22.

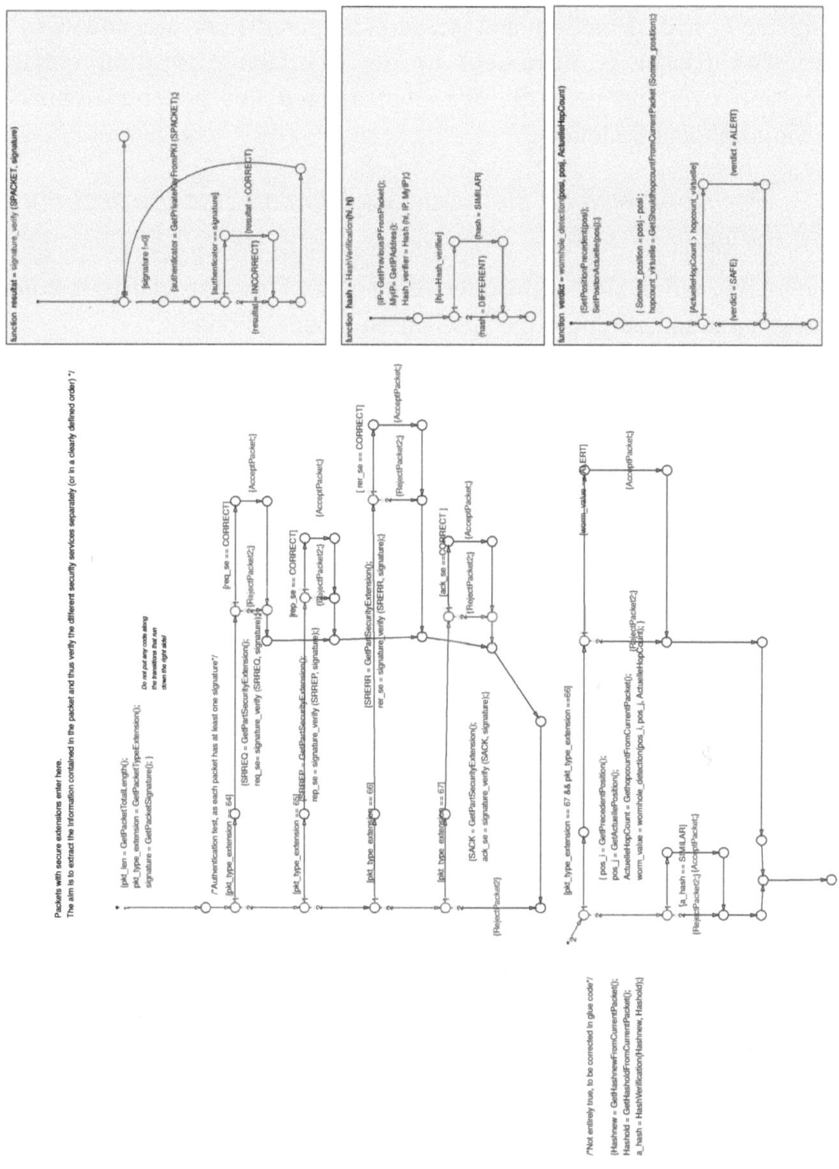

Figure 3.22. *Part of the state transition diagram for the content_extractor block. For a color version of this figure, see www.iste.co.uk/maxa/software.zip*

Additionally, in cases where a packet needs to be retransmitted or if a response packet is sent, an appropriate security header may be added according to the packet type. Mutable and non-mutable messages are authenticated, and detection and prevention parameters are added to the packet. The packet is then sent to the interfacing partition class for transmission over the network. We implemented this generic architecture using Simulink and Stateflow.

3.8.2.3. *Partition between the hardware interface and the security and routing partition*

The final partition class is the interfacing class. This class is used to connect hardware input and output to the application input and output.

This partition class has two complementary aims. The first is to receive IP streams, extract the packets and transmit them to the partition elements managing packet routing and security. The second aim is to create a routing packet by creating streams, collating the different routing and security headers.

Note that the interfacing partition was developed directly in C, unlike the routing and security classes, which were modeled using Simulink and Stateflow. This choice is justified by the fact that it is particularly difficult to access hardware using high-level models. Nevertheless, our selected option remains compatible with our development methodology, as each partition is developed independently. Creating a whole partition manually, writing the code by hand, is in no way incompatible with our model.

3.8.3. *Use of the model-driven approach in developing the SUAP protocol*

The MDD approach proposed here, using the formal verification tools supplied by MathWorks (see Figure 2.1), constitutes both a methodology and an environment for rapid prototyping, from functional modeling to material implementation. Starting with a high-level model, we were able to generate code in C using the Embedded Coder tool described in section 2.5. Before obtaining the C code, the model was checked using formal verification tools (included in MATLAB®) in order to confirm that the system and the generated code behaved in the desired manner. Formal verification methods involve the use of mathematical procedures to find all possible executions and identify any design errors. The advantage of using the tools provided in MathWorks is that verification may be applied to both the model and the code. Models are subjected to formal verification in order to correct design errors, simulating the model in order to validate each partition. The formal verification of code involves static analysis of the code and the use of formal methods to detect different types of errors, including variable overflows, division by zero, unauthorized access to a class variable, and other errors in the execution of the source code. The two following tools were particularly helpful:

– First, the SUAP model was run through the Simulink Design Verifier tool, which applied formal methods to confirm the correct operation of the model. It reviewed each functional block within the model, checking for errors such as integer overflow, dead logic, division by zero and table access violations. For each identified error, the tool indicated the block or Stateflow diagram leading to the error using a set of simulation tests, making it easier to debug the model.

– Simulink Code Inspector was then used to compare the generated code to the source model. This tool systematically examines each block, state diagram and parameter of the model in order to identify their counterparts in the operation, operators and data in the source code. It then generates traceability reports, which may be used to improve the model and which can also be submitted to certification authorities as proof of conformity between the model and the source code.

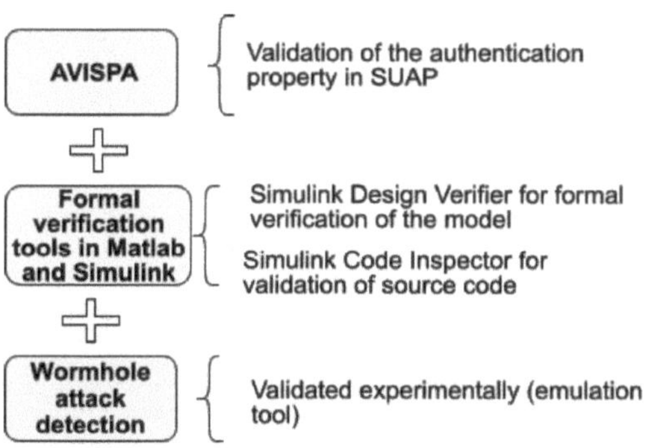

Figure 3.23. *Summary of different validation approaches used in designing the SUAP protocol*

These formal verification tools were used in checking the design of the SUAP protocol. This process is complementary to the verification of security properties carried out using AVISPA. The SUAP detection mechanism was then validated through real experiments, completing the validation phase for the functional requirements of our secure communications protocol (see Figure 3.23).

3.8.4. *Implementation of the SUAP protocol*

The system implementation of the SUAP protocol was obtained by combining the code generated by Embedded Coder (the generator included in MATLAB) with glue code, added by hand.

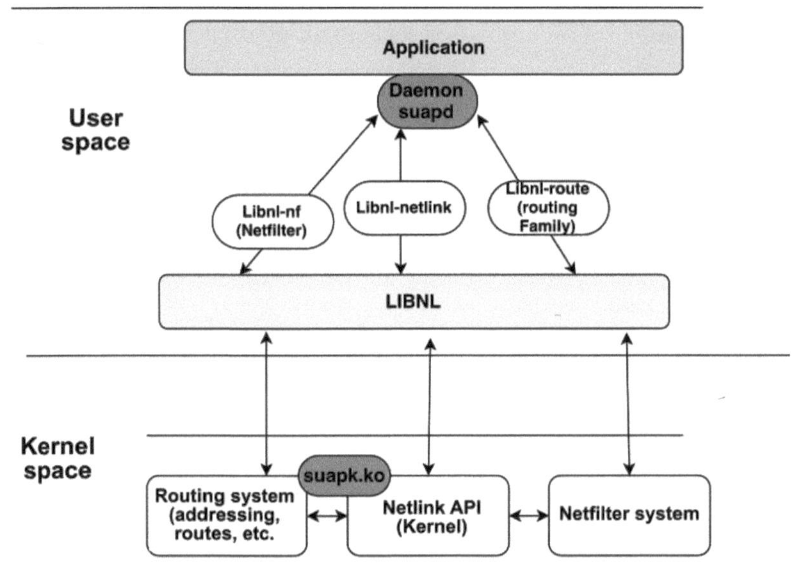

Figure 3.24. *Architecture of the Linux system with SUAP. For a color version of this figure, see www.iste.co.uk/maxa/software.zip*

SUAP is implemented in C (a representation of this implementation is shown in Figure 3.24) and localized in the kernel space (execution of the suapk.ko module) in Linux. It is also characterized by a *daemon suapd* in the user space, which communicates with the kernel module via the libnl [GRA 11] (Netlink Protocol Library). The library allows the kernel to communicate with the user space via a Netlink socket. These communications consist of an exchange of data, enabling, for example, the construction of routing messages and the sending and receiving of data.

libnl also offers methods for managing the netfilter framework. This framework is responsible for managing the network packets traveling through a machine. It supplies system calls that allow the routing protocol to handle packets traveling through network interfaces.

A cross-compilation procedure was used to compile the kernel module obtained in the Linux system with the ARM architecture used by the DT18 drones. The following operations were needed in order to execute the protocol in a DT18 drone:

1) Cross-compilation of the SUAP protocol. This step consists of generating an executable for an architecture other than that used to create the protocol. In our case, we used a cross-compiler for the ARM architecture; the PC we used had instructions in x86. This cross-compilation method saved time during the compilation process (PCs are more powerful than ARM platforms) and allowed us to use more user-friendly development, such as project management programs, which are not available in the target architecture due to the absence (in our case) of a graphical interface.

2) Kernel generation using OpenEmbedded and BitBake [LAU 05]. The SUAP protocol operates in a Linux environment constructed using OpenEmbedded. We therefore need to take account of the executables that make up the SUAP protocol when generating the set of files for use in the ARM system.

3) Configuration of the embedded Linux kernel. This stage is necessary to compensate for missing dependencies in the kernel source files.

4) Unit testing of the SUAP protocol. This involves executing the model on the ARM cards in question, configuring the modem in ad hoc mode.

Note that a key generation application was also created for pairing with the SUAP protocol. As the creation of the key management system lies outside the scope of this study, we decided, in the current version of our protocol, to load the required keys automatically on drone start-up.

3.9. Validation of the SUAP protocol by performance evaluation

In this section, we shall present the validation of the routing and security functions of the SUAP protocol carried out through performance evaluation. This validation complements the formal verification of non-functional requirements carried out using MATLAB and Simulink tools, and the security property verifications carried out using AVISPA. This performance study was carried out, in part, using the hybrid emulation and simulation tool presented in [MAX 15]. This tool is designed to provide a close approximation of real flight conditions for a drone fleet, allowing us to test our protocol locally before evaluating it in a real environment. The protocol was therefore implemented in an operating system similar to that used by the drones in question. Within the tool, we also made use of real drone mobility models. The decision to combine the concepts of simulation and emulation was also motivated by a desire to limit the amount of development work involved in implementing the protocol. Only one version of the code (written in C) is required in this case for implementation in both x86 and ARM architectures. The second element used in the study was a set of real test scenarios, designed to validate the routing and security functions of the SUAP protocol in a real environment. Real tests are valuable in validating the expected functions of an algorithm, developed using a model-driven approach, in a real environment.

We shall begin this section by presenting the experimental framework, followed by the validation of each partition, describing both the network topology used for evaluation and the different metrics used to quantify the network performance of our protocol. We shall then discuss our results.

It is important to note that the performance evaluations presented here consisted of verifying the different functions of our routing protocol. In other terms, they relate to the rapid prototyping methodology described in Chapter 2, enabling us to validate functions which could not be validated using the formal verification tools found in MATLAB, Simulink and AVISPA. In this specific case, we were able to validate the final stage of our methodology (shown in Figure 2.1) concerning the execution of the final binary on the final system.

3.9.1. *Validation of the routing partition*

3.9.1.1. *Test topology for the routing partition*

We began by focusing on validating the routing partition (based on the AODV protocol). We chose to develop a new version of the AODV protocol, based on the rapid prototyping method presented in the previous chapter. This choice stemmed from our desire to facilitate certification of the final product: it is easier to obtain certification using a model-driven design approach than to have existing software code certified (something which is possible, but often costly, as the process is much longer). Our model is described in section 3.8.2.1. We then turned our attention to the validation stage, which involved testing the functions of this partition. To do this, we used a three-node topology, as shown in Figure 3.25.

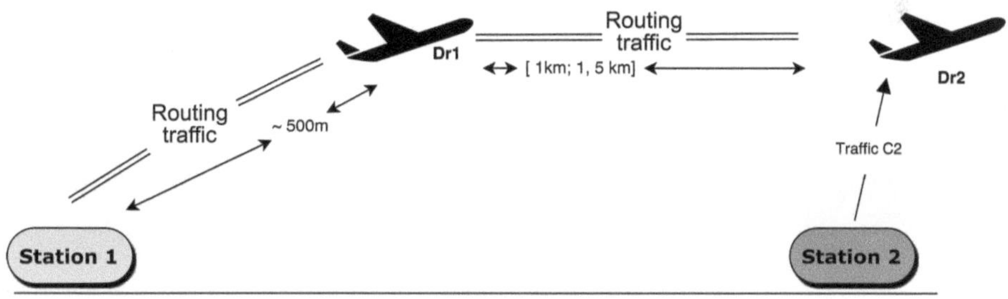

Figure 3.25. *Test topology for the SUAP protocol routing partition. For a color version of this figure, see www.iste.co.uk/maxa/software.zip*

The objective in this case was to transmit video from drone Dr2 to the ground control station via drone Dr1. The exchanged traffic is shown in Table 3.11. A third drone was not needed for validation in this case; three nodes were sufficient to evaluate the performance of the routing element of the SUAP protocol.

The application traffic exchanged between the nodes can be divided into four classes:

1) Tick: signal flow used to inform the drone that it is still in contact with the GCS. Around 100 bytes, sent at a frequency of 1 Hz.

2) Georef: flow used to determine the position of the drone.

3) The actual drone command traffic, used to send commands via a monitoring application. The movement of drone Dr1 is shown in Figure 3.26.

4) Video: traffic emitted in real time from Dr2 (using an embedded camera) to station 1 via drone Dr1.

Type	Source —— Destination	Packet size	Exchange rate
Tick	Station1 —— Dr1 Station2 —— Dr2	64 bytes	1 packet/sec
Georef	Dr1 —— Station1 Dr2 —— Station2	80 bytes	3 packets/sec
Command	Station 1 —— Dr1 Station 2 —— Dr2	80 bytes	1 packet/sec
Video	Dr2 ——— Station1	1400 bytes	25 UDP datagrams/sec width=720,height=576

Table 3.11. *Different flows exchanged during the flight test*

Figure 3.26. *Movement of drone Dr2. For a color version of this figure, see www.iste.co.uk/maxa/software.zip*

3.9.1.2. *Performance metrics*

The following metrics were used to evaluate the performance of the routing protocol:

– Overhead: used to measure the quantity of signaling traffic exchanged between the nodes.

– Route stability: used to measure the duration of uninterrupted connectivity.

– Route re-creation delay: used to measure the routing protocol's reactivity in finding a new route when a connection is lost.

– End-to-end delay.

– Data packet loss rate.

In each case, the values obtained during testing were compared with the evaluation results for the AODV protocol obtained through emulation. These results are summarized in Table 3.12. This comparison allows us to situate the values obtained through real tests and to assess the relevance of our results in relation to a reference implementation of the AODV protocol.

Metric	**AODV protocol**
Overhead (Control packets)	501 kb
Overhead (% Traffic (bytes))	0.034%
Average end-to-end delay	5.32 ms
Loss rate	3.05%
Connectivity	88.5%
Route lifespan	18.53 seconds
Delay in re-establishing a route after connection loss	1.94 ms

Table 3.12. *Summary of results obtained for the AODV protocol: evaluation through emulation*

3.9.1.3. *Results and performance of the routing partition*

3.9.1.3.1. Overhead size

Table 3.13 shows the overhead results we obtained. Looking at these results and comparing them with the performance of the AODV reference protocol, as shown in Table 3.12, we see that the overhead generated by our routing protocol remains within the same order of magnitude as that observed in the emulation tests. This shows us that the traffic generated by the routing protocol does not hinder the transfer of useful data. In this case, the result is explained both by the use of non-acknowledgment for packets and by the fact that established routes are not constantly updated. Once a route between station 1 and drone Dr2 has been established, the protocol no longer attempts to discover the full topology of the network. Each node simply verifies, periodically, the symmetry of the connection with its direct neighbor (one hop) using Hello packets. Our protocol does not generate a significant number of control packets (and thus makes limited demand on the allocated bandwidth).

Overhead	value
Control packets	552 ko
Traffic % (bytes)	0.05%

Table 3.13. *Protocol overhead for a 28 minute test*

3.9.1.3.2. Route stability

The overhead generated by a routing protocol is dependent on the mobility of nodes in the network. The higher the level of node mobility, the more route re-discovery messages will be sent between nodes, indicating route instability.

To evaluate route stability, we simply calculate the number of route request messages sent by the nodes. The total number of route breakdowns during a mission is proportional to the number of route request passages sent during the period in question.

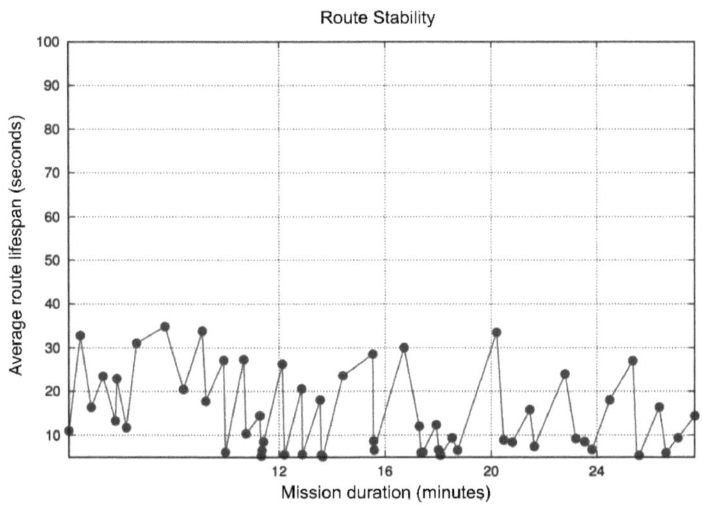

Figure 3.27. *Route stability between nodes GCS 1 and Dr2*

Considering these results (Figure 3.27 and Table 3.14), we see that a route breaks down, on average, once every 14.32 seconds. Table 3.12 gives an average route lifespan of 18 seconds (obtained through emulation and simulation with the same type of mobility observed during our real tests); our implementation thus does not perform as well, although the figures remain within the same order of magnitude. The emulation tool fails to take account of the position of the antenna on each node for each change in the movement model during the mission. The tool we used is not able to simulate losses in attenuation of the signal received by an antenna, and these losses are therefore not quantified.

Metric	Value
Average route lifespan	14.34 s
Maximum route lifespan	34.85 s

Table 3.14. *Route stability between Dr2 and station 1*

The results for this metric are essentially due to the level of node mobility, which leads to route breakdowns. There is also a difference in the radiation and polarization plane of the emission and reception antenna, which may limit the signal. Despite these losses of connection, however, the short time taken to re-establish a route (Figure 3.30), of the order of 2 ms, indicates that these route breakdowns do not have a significant detrimental effect on communications.

In order to illustrate the reasons for these route breakdowns, we have correlated the analysis traces for the network packets with the drone movement log. The traces show that route maintenance and rediscovery packets were sent each time a drone turned (changing direction), for example, during a flight path adjustment in reaction to a position message. One reason for the loss of connection is the change in the plane of the emission antenna of drone Dr1 and of the reception antenna of drone Dr2 (by inclination) at any given moment. More precisely, changes result from the fact that the two drones use different mobility models - for example, drone Dr1 may switch to waypoint mode whilst Dr2 is operating in circular mode. The resulting combination of propagation losses and signal attenuation may result in the temporary loss of a connection.

Furthermore, route outages may also be explained by a variety of parameters intervening in a real situation. When a drone is turning, for example, the inner wing blocks emissions from the onboard modem, located on the underside of the drone. In general terms, the number and range of movements made by the drone make it difficult to maintain communications over long periods.

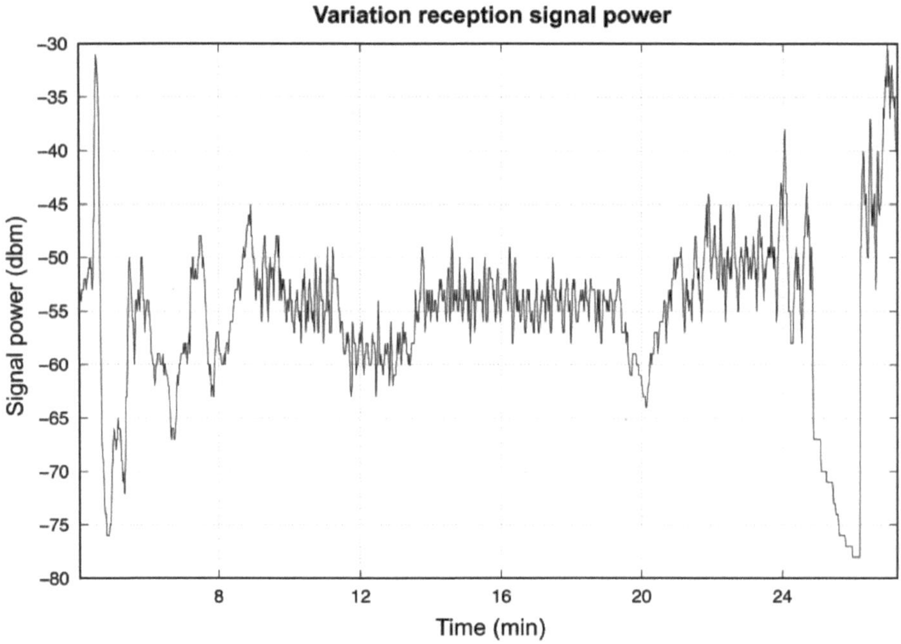

Figure 3.28. *Variation in signal reception power in Dr2*

Figure 3.28 shows that the signal reception power of drone Dr2 oscillates constantly, with significant peaks occurring intermittently. The constant oscillation is due to the unstable movement of the drone during the execution of a specific flight plan; for example, when the pilot sends the command to follow a straight line, the real flight path of the drone will not be completely straight due to inclinations caused by climatic conditions (such as wind speed and direction). Given that the commands sent by the GCS constitute a series of rectilinear, circular or rectangular movements, these elements have an impact on network performance. From Figure 3.29, which shows the correlation between signal power and route lifespan, we see that the delay tends to increase (i.e. service quality improves) as the quality of the signal increases. As the bandwidth allocated to data remains constant, this fluctuation must be due to other physical and topological factors.

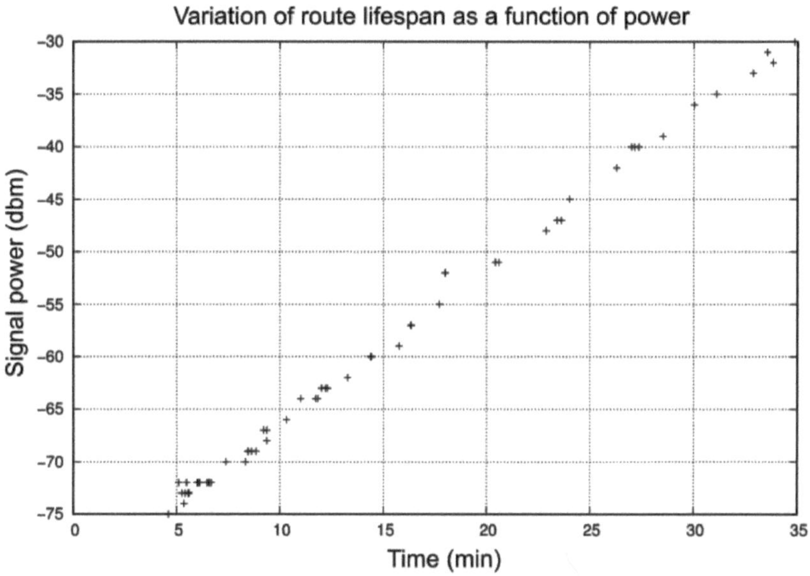

Figure 3.29. *Average stability of a route between Dr2 and the GCS as a function of the signal power of drone Dr2*

3.9.1.3.3. Route re-establishment times

The time taken to re-establish a route following a loss of connection is equal to the time interval between the departure of a request packet (identified by a sequence number) and the arrival of a response packet at the same node (in this case, drone Dr2).

Delay	value
Average delay	2.08 ms
Maximum delay	4.30 ms

Table 3.15. *Time taken to re-establish a route following connection loss*

As we can see, these delays are relatively short (Table 3.15 and Figure 3.30). Comparing these results with the AODV emulation, which gave an average value of 1.94 ms for this metric, our protocol appears to adapt well to topological changes caused by drone movements. This may be explained by the fact that the protocol does not wait for multiple packets to be lost before looking for a new route. The route repair times are thus relatively low for such a dynamic environment. Note also that if a connection between two neighboring nodes (for example, Dr1 and Dr2) is interrupted, maintenance packets are only used to re-establish a route between these two nodes. Here, the speed of node movement has little impact on this value, which tends toward a constant.

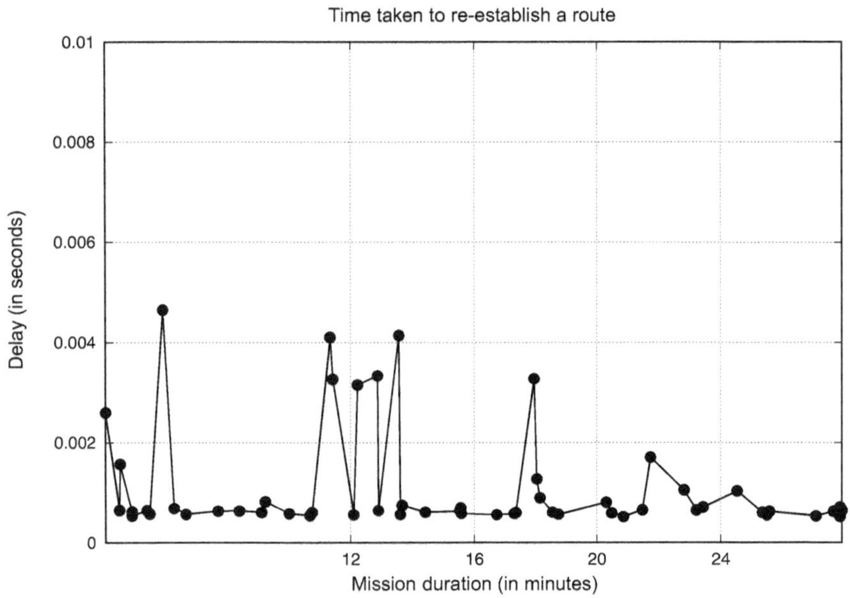

Figure 3.30. *Time taken to re-establish a route following connection loss*

3.9.1.3.4. End-to-end delay

End-to-end delay is the time taken to transfer a packet from an emitter to a final receiver.

Delay	**Value**
Average delay	5.49 ms
Maximum delay	93 ms

Table 3.16. *Average delay between station 1 and Dr2*

The distribution of these delay values is shown in Figure 3.31.

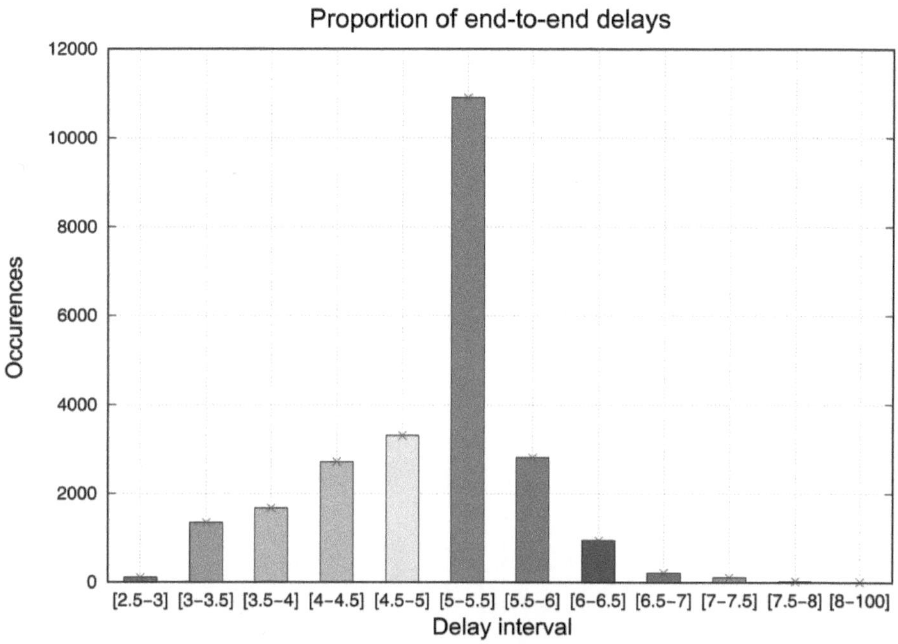

Figure 3.31. *Distribution of end-to-end delay values over the course of the mission. For a color version of this figure, see www.iste.co.uk/maxa/software.zip*

We see that these delays are concentrated in the [5, 5.5] millisecond interval. Comparing this result with Table 3.12, we see that our implementation acts in an equivalent manner to the reference version of AODV, tested via simulations. Considering the low delays observed in both versions, the two protocols do not result in a significant added delay. However, our test results are slightly higher than those obtained through simulation for the AODV protocol. We believe this to be due to the fact that the routing protocol (as modeled) is subject to the introduction of additional parameters encountered in the real-time environment, such as drone speed and a constantly changing network topology.

As the service quality criteria for our data flow essentially require short transfer times, the established hold time is sufficiently brief. This short delay makes it possible to rewrite the video flow acquired by the drone at GCS level.

3.9.1.3.5. Average transfer time for payload data packets

This metric represents the time taken for a video packet to travel from drone Dr2 (source node) to the ground station (destination node).

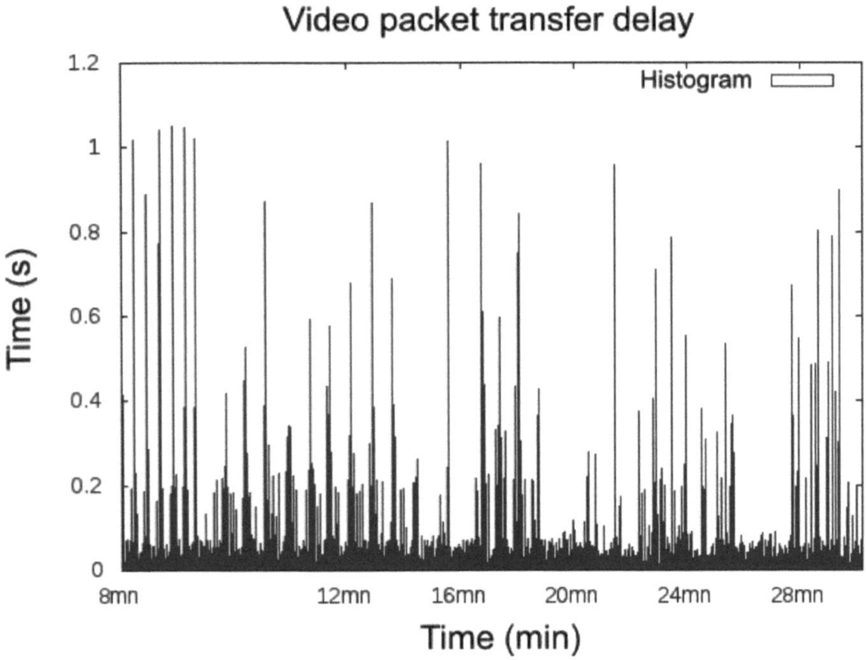

Figure 3.32. *Average payload packet transfer delay*

From Figure 3.32, we see that the delay tends toward a constant, but with peaks which correspond to route interruptions. These peaks are generally due to route loss, and thus to the packet hold time. Based on what we saw above, these route interruptions are mostly due to the node mobility level and to loss of polarization in the antennas. These elements result in connection losses in the network. In this case, data packets are stored in the buffer for a time which depends on the route re-establishment delay.

As the service quality criteria for our data traffic prioritize rapid transmission, with low significance accorded to loss rates, the established hold period is sufficiently short. This low delay value makes it possible to rewrite the video flow acquired by the drone at receiver (GCS) level.

Delay	Value
Average delay	9.15 ms
Maximum delay	1.05 s

Table 3.17. *Average payload packet transfer delay*

3.9.1.3.6. Data packet loss

This metric is designed to measure how many data packets (video traffic) are lost over the course of the mission - in other words, the quantity of packets lost between the moment a connection is lost and the moment it is re-established. The evolution of loss rates over time is shown in Figure 3.33. We see that, over the course of the mission, the loss rate does not exceed 4%. In our emulation tests, we obtained an average value of 3.05%, compared with 3.51% for our real tests.

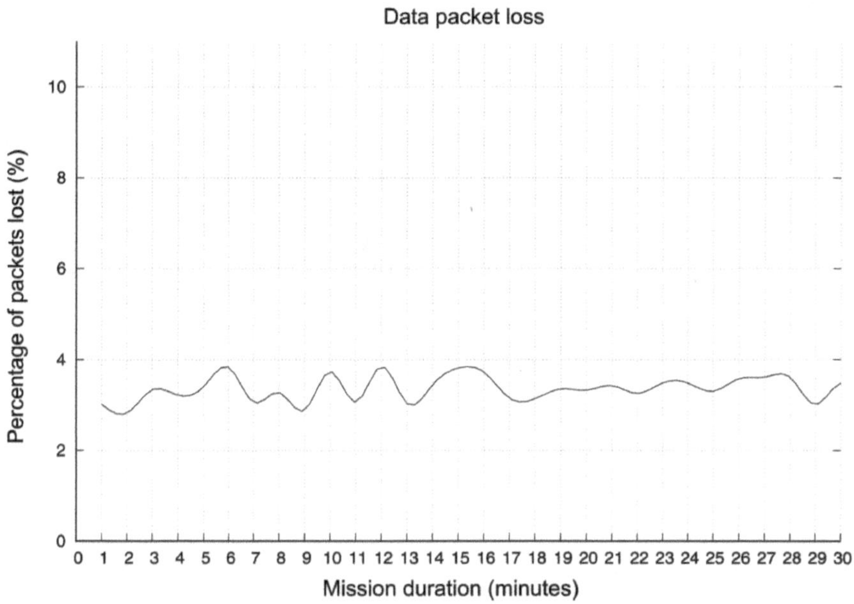

Figure 3.33. *Measured loss rates. For a color version of this figure, see www.iste.co.uk/maxa/software.zip*

These two final results (loss rate and video packet transfer delay) allow us to validate the performance of our implementation of the AODV protocol. AODV responds to the quality of service criteria identified in our initial specification, providing the service for which it was designed and implemented.

3.9.1.4. *Validation of the routing partition: discussion*

So far in this section, we have described the validation of the routing partition of the SUAP protocol. The results obtained are acceptable with regard to the requirements of video traffic. Table 3.18 shows the results obtained via simulation using AODV alongside those obtained through real testing. We have considered the quantity of signaling traffic generated over the course of a mission: this quantity only takes up a small proportion of the available bandwidth, as our protocol is reactive and only generates route discovery packets on an as-needed basis. The end-to-end delay and the route re-establishment delay are generally low, due to the low node density. This means that the speed of movement of the nodes is of relatively little importance. The latter results should be considered in relation to the performance criteria obtained through simulation, comparing the different routing protocols used in the ad hoc domain (notably an existing version of the AODV protocol). Our implementation of AODV, created using a model-driven rapid prototyping method, produces satisfactory results.

Parameter	Emulation result for AODV	Real test using the modeled version of the AODV protocol
Loss rate	3.05%	3.50%
Average end-to-end delay	5.32 ms	5.49 ms
Average route re-establishment delay	1.94 ms	2.08 ms
Route stability	18.53 s	14.34 s
Overhead	501 kb (0.034%)	552 kb (0.05%)

Table 3.18. *Summary of results obtained through emulation and real testing*

Following these tests, we were able to continue our MDD development cycle, a process which goes from specification all the way to launch. Our implementation may be seen to offer high adaptability with regard to the specificities of an ad hoc drone network (node speeds, mobility levels, propagation models, number of nodes and dynamic topology). In the next section, we shall consider the validation of the security partition of the SUAP protocol (covering authentication and integrity).

3.9.2. *Validation of the security functions of the SUAP protocol*

In this section, we shall discuss the validation of the message authentication and integrity mechanism used in the SUAP protocol. This validation was carried out both through emulation and in a real-world environment. We shall begin by focusing on the emulation study.

3.9.2.1. *Validation of the security partition (message authentication) in an emulated environment*

Our aim was to valuate those security properties of the SUAP protocol, which cannot be verified by AVISPA or by the internal verification tools found in MathWorks' Simulink and Stateflow. To do this, we considered the topology shown in Figure 3.34, to simulate a single black hole attacker, and, in Figure 3.35, for two black hole attackers. We shall evaluate the security performances of the SUAP protocol for both configurations.

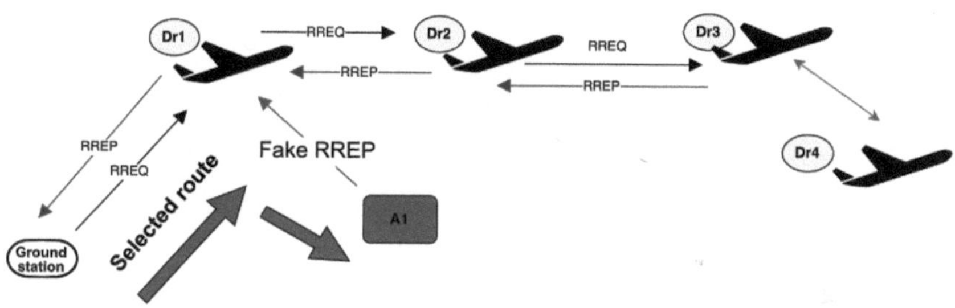

Figure 3.34. *Illustration of the black hole attack. For a color version of this figure, see www.iste.co.uk/maxa/software.zip*

These two forms of the black hole attack allow us to validate the message authentication and integrity mechanisms used in SUAP. Each packet generated by any node, whether an attacker or a genuine node, must be authenticated before being analyzed and transmitted as necessary. To assist understanding of this section, a description of the black hole attack is given below, based on the two test configurations used here.

In the first configuration (shown in Figure 3.34), we have four genuine nodes. The GCS begins by searching for a route toward drone Dr4 in order to send data. The packet is broadcast over the network until it reaches drone Dr3, which knows a route to node Dr4. Node Dr3 thus sends a response packet indicating that it knows a route to the desired destination. At the same time, the attacking node A1 creates a fake RREP packet, which it sends to node Dr1, with a better metric (for example, the hop count to reach the destination in question) than that found in the genuine RREP packet generated by node Dr2.

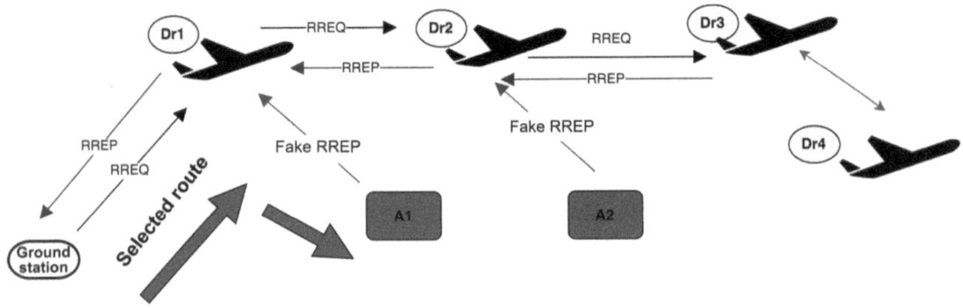

Figure 3.35. *Black hole attack with two attackers. For a color version of this figure, see www.iste.co.uk/maxa/software.zip*

If packets are not authenticated, the GCS will transmit data toward the attacker. To implement this attack, we applied a version of the black hole strategy using node A1, which simply rejects all data packets it receives. Note that another, more sophisticated variant also exists, in the form of the gray hole attack; in this case, only certain packets are rejected. Nevertheless, the original, simpler form of the black hole attack is sufficient to evaluate our security mechanism. In this configuration, node Dr4 will never receive the application data which it is intended to obtain.

In the second configuration, shown in Figure 3.35, a second attacker has been added. The two attack nodes are mobile within the network. This mobility element is interesting in that it allows us to observe the ways in which attackers may interact directly with genuine nodes, and thus to measure the performance of our protocol as a function of the number of attackers. Given the low network density, a simulation with non-mobile attackers would not be sufficient for effective validation of our authentication mechanisms.

3.9.2.1.1. Performance metrics

The following metrics were used to assess the network performance of the SUAP protocol in case of attack:

– delivery rate, measuring the number of packets received by a destination in comparison to the number of sent packets. This metric gives us an idea of the capacity of the routing protocol to maintain a stable route for exchanging data packets in spite of the presence of attackers;

– end-to-end delay between source and destination nodes;

– overhead of the security mechanisms, to evaluate the supplementary traffic created by the use of these mechanisms. This additional cost relates to both the authentication and integrity mechanisms, and allows us to assess whether or not the mechanisms have a detrimental effect on network performance;

– connectivity between different network nodes. This measurement allows us to quantify route stability in the presence of an attacker.

Note that all of the results presented here were obtained over several repetitions of the same simulated scenario, providing us with an average value and guaranteeing reproducibility. We also compared the SUAP protocol with the non-secure version (i.e. the SUAP routing partition) validated above. In this case, we have used the term AODV to denote the routing partition of the SUAP protocol.

3.9.2.1.2. Packet delivery rate

Figure 3.36 shows the packet delivery rates for the SUAP and AODV protocols in the presence of one and two black hole attackers. The results demonstrate a significant difference in both cases. There are several reasons for this:

1) First, in the AODV protocol, the majority of data packets do not reach their intended destination, as they are rejected by the black hole attacker(s).

2) Whilst the attackers eliminate data packets, the delivery rate is not zero, as the real mobility models used in the simulator mean that network nodes, including attackers, may find themselves out of range of the genuine nodes. If attackers are out of range of the network for a period of time, data packets will

be able to reach their destination. As soon as the attackers re-enter the network, packet losses resume.

3) It is also important to note that with two attackers, the data delivery rate is low, as the black hole nodes are mobile and interact with moving drones. The low network density means that the number of attackers has a strong effect on the delivery rate.

4) In the case of the SUAP protocol, the number of attackers has a limited effect on the packet delivery rate. Each sent packet is authenticated by the authentication and integrity mechanism. Fake packets are thus ignored, and data packets are only transmitted along genuine routes. However, the result does not reach 100% due to node mobility, which can cause temporary breakdowns in connection.

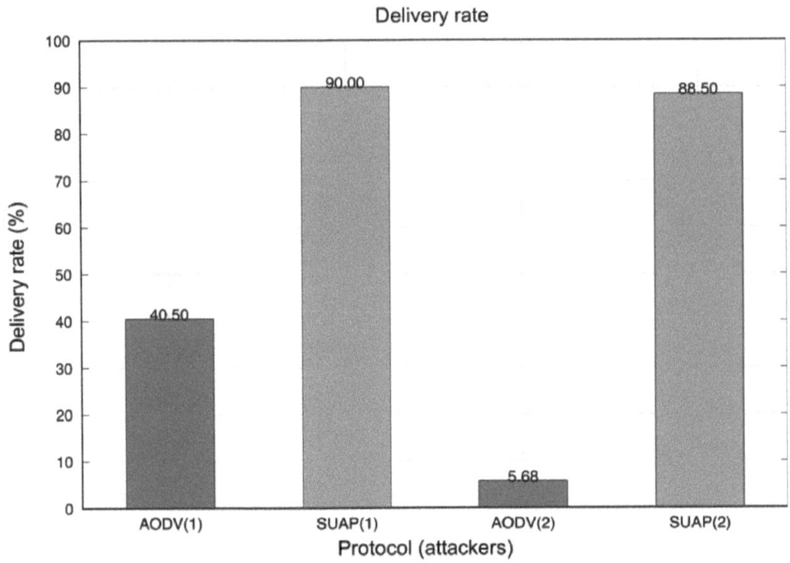

Figure 3.36. *Delivery rate for both protocols with one and two black hole attackers. For a color version of this figure, see www.iste.co.uk/maxa/software.zip*

3.9.2.1.3. End-to-end delay

The end-to-end delay between the GCS and drone Dr4 is shown in Figure 3.37. This delay tends toward a constant for both protocols. Here, note that the focus is not on the end-to-end delay for AODV routing packets, but rather for secure routing. We see that the delay for SUAP is slightly longer than that observed for the normal routing protocol. This slight increase is justified by the signature and signature verification times. These values are acceptable in the context of the real-time application in question. Packet authentication is thus achieved without exceeding the maximum acceptable value for this metric; for the video flow transmitted in this case and for our established specification, a delay of up to 10 ms is allowed.

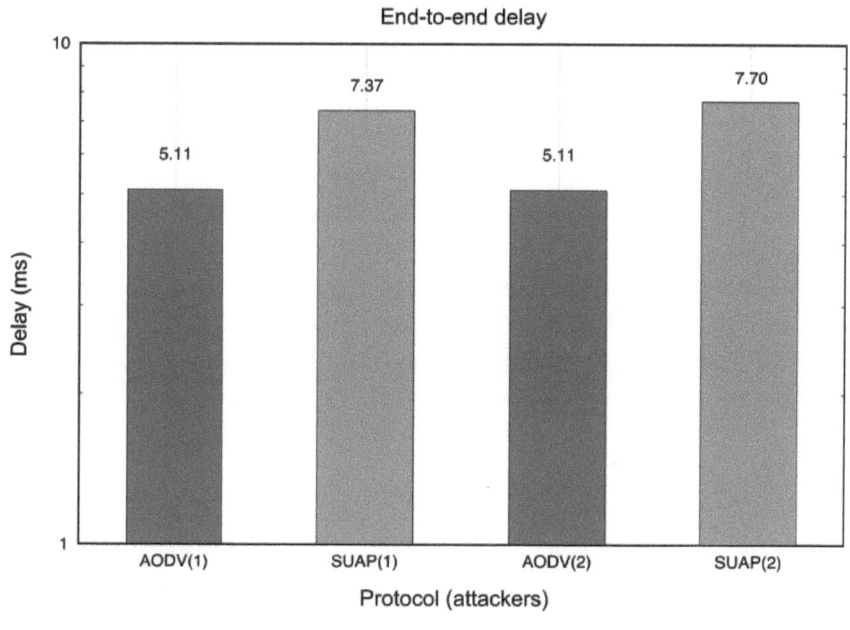

Figure 3.37. *End-to-end delay for each protocol with different numbers of attackers. For a color version of this figure, see www.iste.co.uk/maxa/software.zip*

It is important to note that these values are dependent on the chosen cryptographic algorithms, and notably, on the size of the keys. Our implementation uses RSA-512 for the signature and SHA-256 for hashing. These algorithms were selected on the basis of network performance, as recommended in [ZAP 08].

3.9.2.1.4. Overhead

Table 3.19 shows the cost of security packets in relation to normal signaling traffic. This parameter is important in terms of the bandwidth available for applications. The lower the cost of security services, the more space is left for applications within the communications medium.

	AODV	SUAP
Signaling packets	501 ko	2903 ko
% Traffic (bytes)	0.034%	0.46%
% Security traffic in relation to routing traffic (bytes)		45%

Table 3.19. *Overhead for one hour of testing*

From these results, we see that the SUAP protocol makes higher demands on the available bandwidth than AODV due to the use of security headers. However, this addition represents only 0.46% of the total exchanged traffic. This rate is acceptable given the security functions which it contributes to the final application. Moreover, considered in light of the presence of malicious nodes, this overhead is strictly necessary; not using a security mechanism would result in a very low packet delivery rate (see section 3.9.2.1.2), disrupting the video surveillance mission.

3.9.2.1.5. Connectivity

Our connectivity results are shown in Figure 3.38, which indicates the percentage of route connectivity in the presence of attackers for both protocols. These results indicate that the AODV protocol is strongly affected by black hole attacks. The malicious node A1 is present along the route used during the majority of communications. This results in a network node connectivity level of approximately 10%. This result tends toward an even lower value in the presence of a black hole attacker. Using the SUAP protocol, the genuine route is maintained at 88% of the time. This indicates that the authentication and integrity mechanisms used here are able to maintain network integrity, avoiding the malicious node. This value of 88% is static, and is not proportional to the number of attackers in the network. A 100% connectivity level cannot be reached as node mobility affects the connectivity between neighboring nodes.

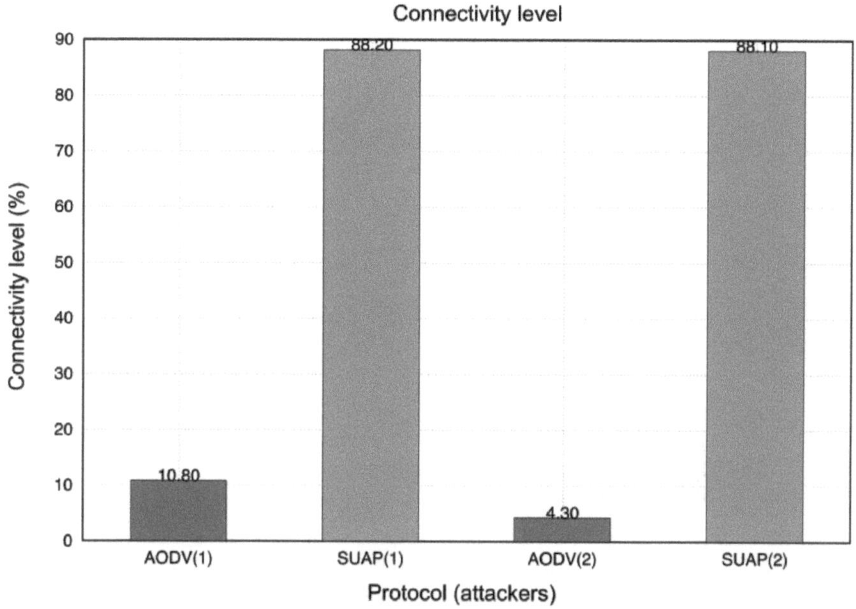

Figure 3.38. *Connectivity results for one hour of testing. For a color version of this figure, see www.iste.co.uk/maxa/software.zip*

Furthermore, comparing the results obtained for this metric with the results from the emulation tests of AODV and the real tests of the routing partition, we obtain a result within the same order of magnitude (88%). This indicates that the security bricks do not result in a significant reduction in connectivity in the UAANET network in case of attack.

3.9.2.2. *Validation of the security partition (message authentication) in a real environment*

The topology used here is shown in Figure 3.39. It is made up of four nodes: three DT18 drones and a ground control station. We added a second ground station, acting as a black hole attacker. For the purposes of this test, we needed to place an attacker near the emitting node Dr3 to verify whether the genuine route through nodes Dr1 and Dr2 would be chosen throughout the test period. We therefore reduced the power of the modem antenna to give us a range less than or equal to 500 m. Station 4 (the black hole attacker) has a maximum emission power of up to 1.5 km. In other terms, the attacker has the capacity to contact all of the nodes in the network, notably the most distant drone. This topology allows us to verify whether or not the SUAP protocol is able to defend against black hole attacks in real conditions.

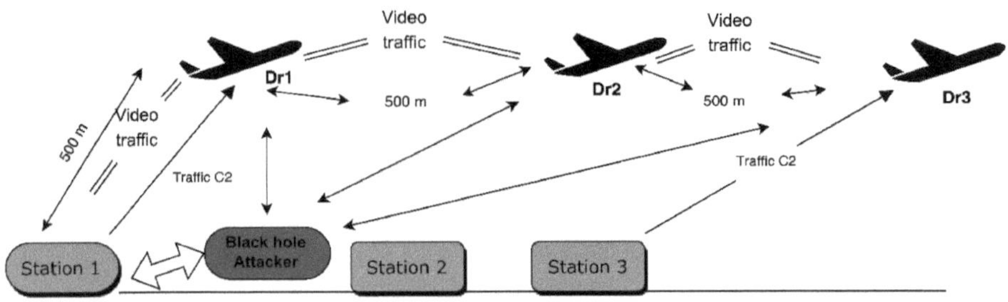

Figure 3.39. *Topology used to validate the security partition in a real environment. For a color version of this figure, see www.iste.co.uk/maxa/software.zip*

It is important to note that other equipment could have been used to play the part of the attacker. The choice of a ground station avoided the need to implement the black hole attack code in a different machine architecture. The previous configurations remained unchanged. The traffic exchanged between nodes is summarized in Table 3.20.

Type	Source — Destination	Packet size	Exchange rate
Tick	Station1 — Dr1 Station2 — Dr2 Station3 — Dr3	64 bytes	1 packet/sec
Georef	Dr1 — Station1 Dr2 — Station2 Dr3 — Station3	80 bytes	3 packets/sec
Command	Station 1 — Dr1 Station 2 — Dr2 Station 3 — Dr3	80 bytes	1 packet/sec
Video	Dr3 — Station1	1400 bytes	25 UDP datagrams/sec width=720,height=576

Table 3.20. *Different flows exchanged during the flight test*

Figure 3.40 shows the contents of the TTL (Time to Live) field for each of the nodes. It indicates that between 7 and 23 minutes into the test, the three drones were aligned in-flight and communicating in ad hoc mode. We have therefore chosen to study the behavior of the SUAP protocol over this period.

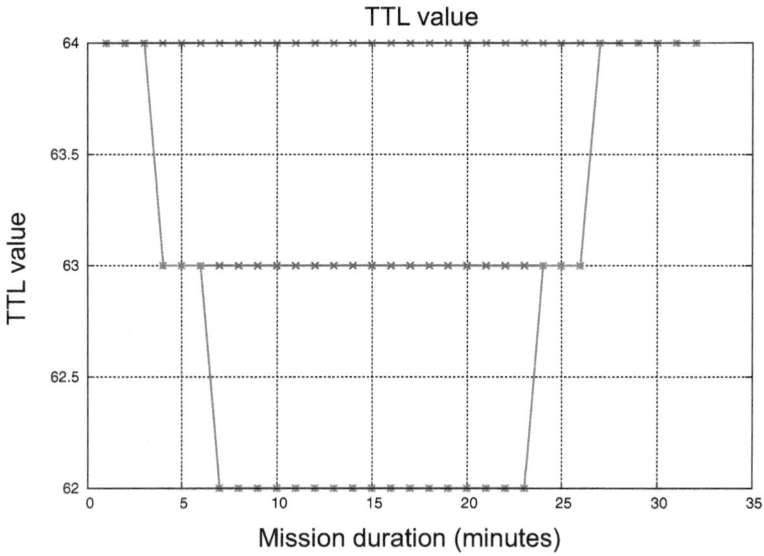

Figure 3.40. *TTL values for drones Dr1 (ttl=62), Dr2 (ttl=63) and Dr3 (ttl=64). For a color version of this figure, see www.iste.co.uk/maxa/software.zip*

3.9.2.2.1. Packet loss rate

The packet loss rate is shown in Table 3.21. As we can see, losses are concentrated around the connection between the three drones. The mobility and different movement patterns of the drones during the test resulted in connections being lost and re-established over the course of the test. The total observed loss rate is 5.57% for data packets sent from node Dr3 to node station 1. This result is acceptable compared with the results obtained through emulation.

Source	Destination	loss rate
Station 1	Dr1	0.53
Dr1	Dr2	2.32
Dr2	Dr3	2.70
Total (Station 1 — Dr3)		5.57

Table 3.21. *Loss rates (%) measured between each node*

3.9.2.2.2. Overhead

The results for the overhead are shown in Table 3.22, and are also similar to those obtained through emulation. The reactive nature of the routing protocol means that few traffic signaling packets are exchanged within the network. The protocol only acts if a route creation request is emitted. The low density of nodes means that drone movement speed has no direct impact on this value.

Source	Destination	Size of control packets (in bytes)	Percentage of total traffic
Station 1	Dr1	89 760	0.075%
Dr1	Dr2	153 653	0.093%
Dr2	Dr3	153 321	0.0927%
Total			0.270%

Table 3.22. *Overhead for the 16-minute test period*

3.9.2.2.3. Average route lifespan

Our results for the average lifespan of routes between station 1 and Dr3 are shown in Table 3.23 and Figure 3.41. These results may be compared with Figure 3.27 and Table 3.14, which illustrate route stability using the AODV protocol (with no attackers in the network). We see that the average lifespan of a route remains within the same range with and without security. This indicates that our routing protocol behaves in the same way as AODV in performance terms, and that the addition of security mechanisms does not have a detrimental effect on network performance, whilst providing defense against attacks in this configuration.

Delay	Value
Average delay	15.15 s
Maximum delay	20.44 s

Table 3.23. *Route stability in the presence of a black hole attacker*

The delay involved in re-establishing a route in response to a loss of connectivity between one or more nodes, causing route loss, is shown in Table 3.24. We obtained an average delay value of 2.23 milliseconds, which is sufficient to compensate for frequent connectivity losses in the network. This result is in the same range as the route re-establishment value obtained for the AODV protocol through real testing (2.23 ms compared with 2.08 ms).

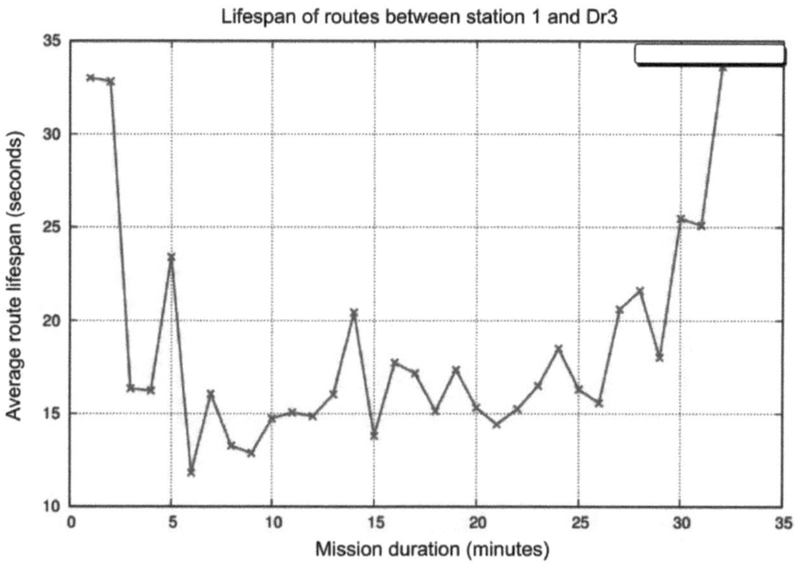

Figure 3.41. *Route stability in the presence of a black hole attacker. For a color version of this figure, see www.iste.co.uk/maxa/software.zip*

Delay	value
Average delay	2.23 ms
Maximum delay	5.79 ms

Table 3.24. *Route re-establishment delay following a loss of connection*

3.9.2.2.4. Percentage of routes established between station 1 and Dr3

The purpose of this metric is to evaluate the number of route creations passing through the legitimate route (made up of nodes Dr3, Dr2, Dr1 and station 1) and passing through the black hole attacker. The aim here is to verify whether packets originating from the malicious node lead to the creation of a route. Our results are shown in Figure 3.42. The red line represents packets from drone Dr2, and the green line represents packets sent by the attacker. To calculate this metric, we filtered the IP address of the attacker (using the measured trace) in node Dr3, checking to see whether or not it was accepted as a route initiator.

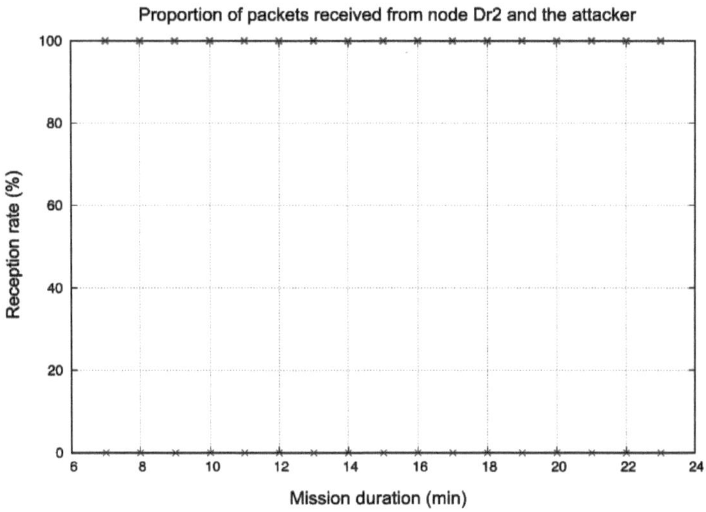

Figure 3.42. *Percentage of control packets leading to the creation of a route from Dr2 and the attacker, measured at the interface with node Dr3. For a color version of this figure, see www.iste.co.uk/maxa/software.zip*

From this figure, we see that 100% of the packets leading to route creation originated from node Dr2, the genuine neighboring node. Dr3 did not process any unauthenticated packets from the malicious node. This indicates that the algorithms used to authenticate messages operate correctly.

3.9.2.2.5. End-to-end delay

The end-to-end delay for each traffic type is shown in Table 3.25. This value is generally low (under 10 ms on average) and does not have a negative impact on real-time traffic exchanges. However, it is higher than the result obtained for the AODV routing protocol. This is due to the non-negligible time taken by the LIBGCRYPT cryptography library[7], used to create and verify signatures and hashes. Figure 3.43 shows the evolution of the end-to-end delay over the course of the mission, illustrating the proportion of packets with a delay of over 10 ms. We see that our results are concentrated in the interval [7, 7.5]. Delays of over 100 ms occurred only on seven occasions during the mission.

---

7 For more information, see https://www.gnupg.org/software/libgcrypt/index.html

End-to-end delay for signalling traffic exchanges	Value	End-to-end delay for payload traffic exchanges	Value
Average delay	7.43 ms	Average delay	9.2 ms
Maximum delay	100 ms	Maximum delay	104 ms

Table 3.25. *End-to-end delay for control packets and real-time traffic*

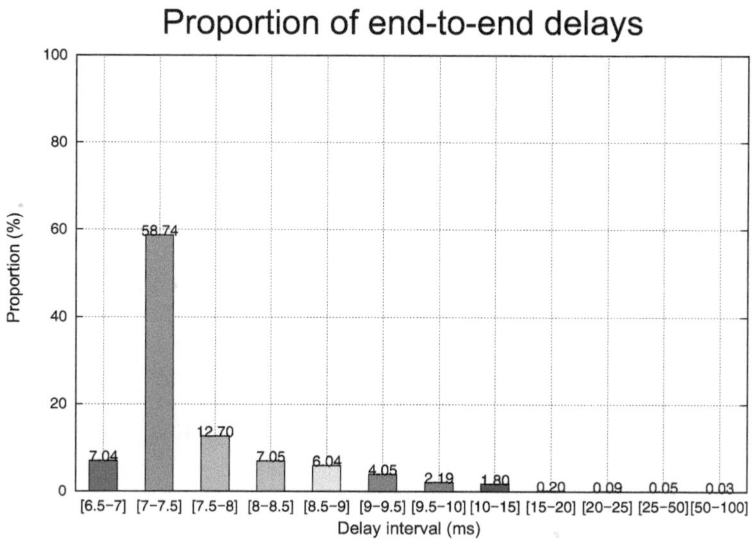

Figure 3.43. *Distribution of end-to-end delay values over the course of the mission. For a color version of this figure, see www.iste.co.uk/maxa/software.zip*

3.9.2.2.6. Evaluation of the AODV protocol in a real environment with a black hole attacker

To gain a clearer understanding of the results we obtained for our message authentication mechanisms, we tested the AODV protocol in a real environment with a black hole attacker. This test was carried out using the same configuration used for SUAP, as shown in Table 3.20 and Figure 3.39. Our results are summarized in Table 3.26. These results are shown alongside those obtained for the SUAP protocol, as seen above. Note that only those metrics for which AODV and SUAP produced significantly different results are shown here.

Parameter	Real test of the AODV protocol with a black hole attacker	Real test of the SUAP protocol with a black hole attacker
Loss rate	98.20%	5.57%
Percentage of routes established between Station 1 and Dr3	1.76%	100%
Overhead	210 kb	397 kb

Table 3.26. *Summary of results obtained for real environment evaluations of the AODV protocol with a black hole attacker*

Note that drone mobility was not the same in this case as in our evaluation of the SUAP protocol, due to the difficulty involved in reproducing exactly the same movement patterns in a real environment. However, our results clearly show the weakness of the AODV protocol to black hole attacks in a real environment. These results may be explained as follows:

– For the loss rate, the majority of data packets are lost in this case, because each time a route request is sent, the black hole attacker intervenes and emits fake packets, leading to erroneous routes.

– The percentage of routes established between station 1 and node Dr3 is only 1.7% for the whole mission, as the vast majority of established routes included the black hole attacker.

– The overhead value obtained for the AODV protocol is smaller than that for the SUAP protocol (see the results presented in Table 3.22). The additional traffic results from the security mechanisms included in the latter protocol. However, it is not sufficiently high to have a negative impact on data exchange, as we saw above (see the end-to-end delay results presented in Table 3.25).

3.9.2.3. *Discussion*

In this section, we have presented our validation study for the message authentication partition of the SUAP protocol. The results obtained through emulation and in a real environment show that the SUAP protocol offers a robust defense against black hole attacks. We focused on black hole attacks as they allow us to check whether each node in the network is able to verify the security level of messages before processing them. The black hole attack can be seen to have no direct influence on the performance of the SUAP protocol, whilst having a significant effect on the AODV routing protocol. We may thus state that our implementation, developed using a model-driven approach, demonstrates the high-level behaviors set out in the specification. These first results are positive but not sufficient to validate all of the security functions of the SUAP protocol. In the following section, therefore, we shall consider the wormhole detection mechanism proposed earlier in this chapter, which constitutes one of the main components of our security architecture.

3.9.3. *Validation of the wormhole detection mechanism*

In this section, we shall analyze the precision of wormhole attack detection offered by the SUAP protocol. This validation study was carried out by emulation alone. To validate the partition, we considered a topology made up of five genuine nodes and two attackers, as shown in Figure 3.44. We used emulation for this phase of validation due to limited access to mobile nodes and to embedded systems compatible with the DT18's communications systems, which meant that we did not have the means to envisage a real-environment validation approach. The possibility of carrying out real-world tests of this nature will be discussed in the Conclusion.

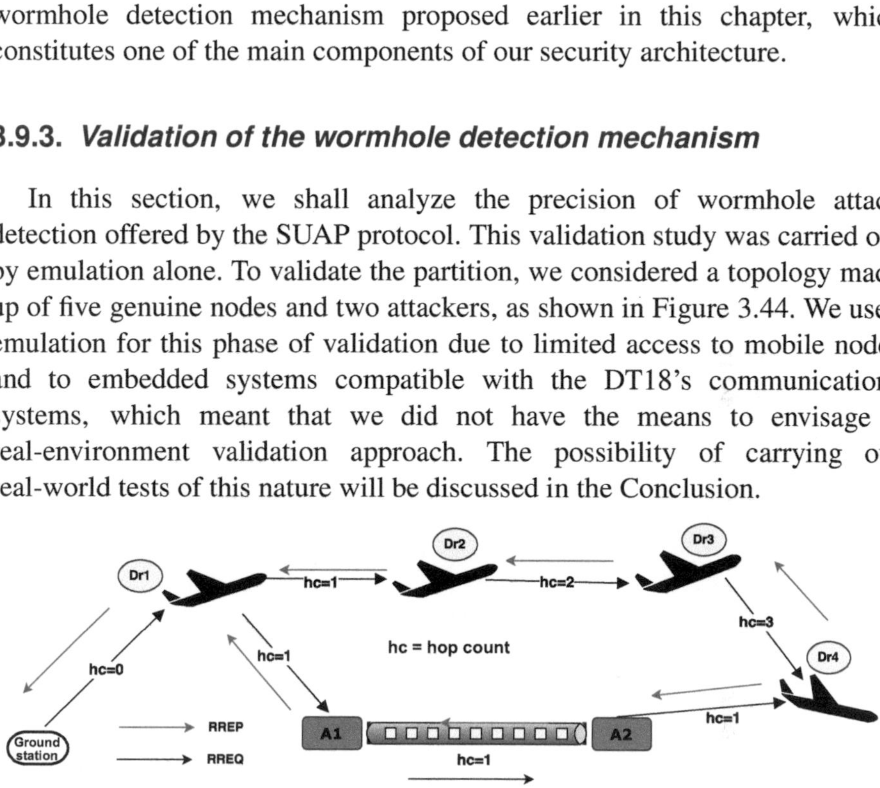

Figure 3.44. *Validation topology for the wormhole attack. For a color version of this figure, see www.iste.co.uk/maxa/software.zip*

The parameters used in our scenario are shown in Table 3.27.

We compared the SUAP and AODV protocols to study the influence of the attack. To do this, we verified that the route passing through the tunnel between A1 and A2 was not used for data transmission. We considered two metrics in assessing the relevance of our mechanism:

1) First, we measured the quantity of data (sent from drone Dr4) received by the ground station (traveling over the two alternative routes). This allowed

us to compare the quantity of data passing through the wormhole and the quantity of data passing through a genuine connection.

2) Second, we evaluated the number of established routes passing through the wormhole and the number of routes passing through the legitimate connection.

Parameters	Value
Number of genuine nodes	5 (4 drones and 1 ground station)
Mobility	Real mobility replay
Routing protocol	SUAP and AODV
MAC protocol	802.11
Radio range	100 m
Simulation duration	600 s
Channel capacity	54 Mbit/s

Table 3.27. *Evaluation parameters used in validating the wormhole attack detection mechanism*

Type	Source→Destination	Packet size	Frequency
Tick	1→2,1→3,1→4	64 bytes	1.0 packets/s
Georef	2→1,3→1, 4→1	64 bytes	1.8 packets/s
Command	1→2,1→3, 1→4	64 bytes	0.034 packets/s
Video	2→1,3→1,4→1	4 Mbit/s	

Table 3.28. *Generated traffic: 1=Dr1, 2=Dr2, 3=Dr3, 4=Dr4*

These two metrics are complementary, and sufficient to demonstrate the relevance of our security mechanism in defending against wormhole attacks.

For validation purposes, we decided to simulate two stationary attackers, with A1 located close to the ground control station and A2 close to the most distant node, Dr4. In this way, the attackers were located near the target nodes, making it easier to evaluate the impact of the attack.

Tables 3.29 and 3.30 show the quantity of information exchanged by genuine nodes and by attackers, respectively. We see that, using the SUAP protocol, the wormhole tunnel only transmits routing information for the target nodes. The fact that attackers receive routing packets is not important, as these messages do not contain any secret information relating to the mission.

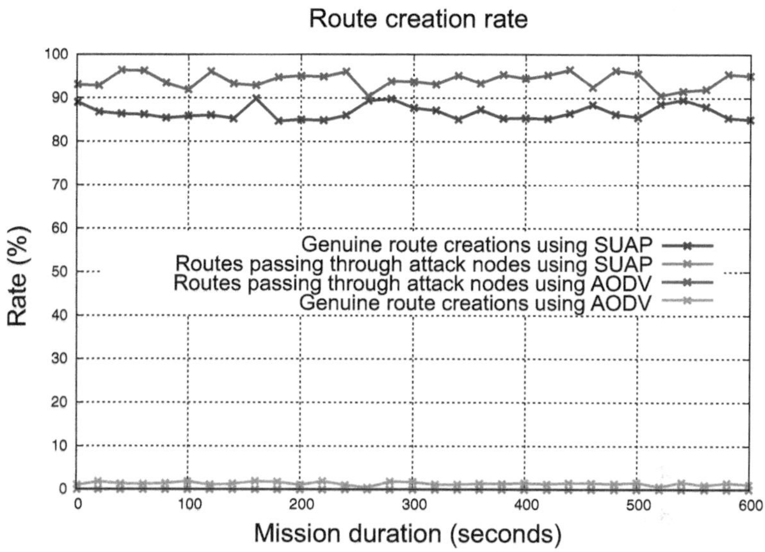

Figure 3.45. *Creation rate for genuine and attack nodes. For a color version of this figure, see www.iste.co.uk/maxa/software.zip*

Source	Quantity of signaling traffic generated	Video traffic generated	Destination	Quantity of signaling traffic received	Quantity of data traffic received
Dr4	146783 bytes	23 02 289 kb	Dr3	146783 bytes	23 02 111 kb
Dr3	140782 bytes	23 02 111 kb	Dr2	140782 bytes	23 00 011 kb
Dr2	142345 bytes	23 00 011 kb	Dr1	142345 bytes	22 98 245 kb
Dr1	130011 bytes	22 98 245 kb	GCS	140011 bytes	22 97 988 kb
Dr4	146783 bytes	23 02 289 kb	A1	146783 bytes	**0 bytes**
A1	146783 bytes	**0 bytes**	A2	146783 bytes	**0 bytes**

Table 3.29. *Summary of data sent and received in communications between Dr4 and the ground control station using the SUAP protocol*

We also see that no video information is transmitted between A1 and A2. Here, detection is carried out by the ground station, which identifies an anomaly between its relative distance from node Dr4 and the hop count in the neighborhood discovery packet received from that node. In this case, Hello packets are protected by a geographical leash algorithm that compares the hop count and the relative distance between neighbors. Request and response packets are protected by a hashing function. Our results indicate that SUAP is effective in choosing a genuine route, despite routes with better metrics being offered through the wormhole.

Comparing these results with Table 3.30, which indicates the quantity of data received by the two attackers using the AODV protocol, we see that the wormhole succeeds in transmitting video from node Dr4. In this case, the other genuine nodes only exchange signaling traffic, receiving only a tiny part of the useful data (0.18% of the total traffic). This is explained by the fact that the route through the malicious nodes is always chosen as it offers a better hop count metric. The value of 0.18% is due to the fact that node Dr4 may, on occasion, be out of range of A2 due to its mobility, and in this case, the genuine route may be used for video transmission over a short period of time, corresponding to 0.18% of the total quantity.

Figure 3.46 shows our results for data delivery rates over both routes (genuine and wormhole) as presented in Tables 3.29 and 3.30. We see that using the SUAP protocol, only the genuine route is used to transmit data packets. This leads us to conclude that the SUAP protocol is robust against wormhole attacks.

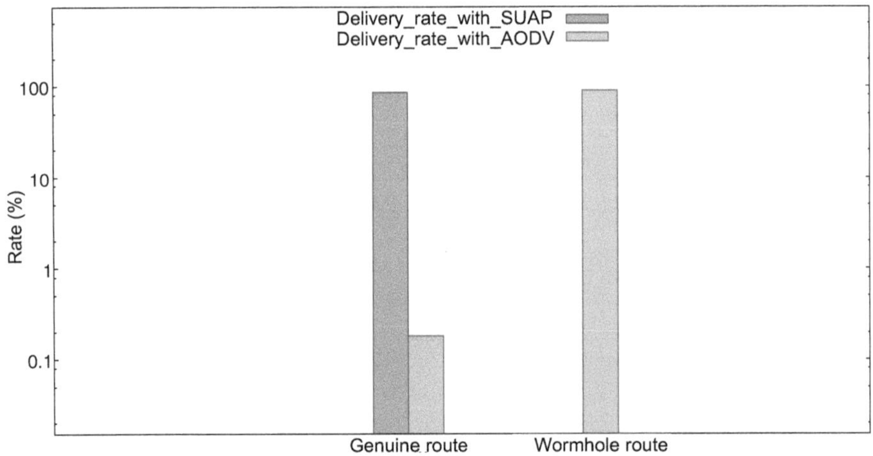

Figure 3.46. *Packet delivery rate in the context of a wormhole attack. For a color version of this figure, see www.iste.co.uk/maxa/software.zip*

Source	Quantity of signaling traffic generated	Video traffic generated	Destination	Quantity of signaling traffic received	Quantity of data traffic received
Dr4	146783 bytes	23 02 289 kb	Dr3	146783 bytes	13450 kb
Dr3	140782 bytes	13450 kb	Dr2	140782 bytes	13450 kb
Dr2	142345 bytes	13450 kb	Dr1	142345 bytes	13450 kb
Dr1	130011 bytes	13450 kb	GCS	140011 bytes	13450 kb
Dr4	146783 bytes	23 02 289 kb	A1	146783 bytes	22 98 839 kb
A1	146783 bytes	22 98 839 kb	A2	146783 bytes	22 98 839 kb

Table 3.30. *Summary of data sent and received in communications between Dr4 and the ground control station using the AODV protocol (no security)*

To improve our understanding of these results, we measured the route creation rate along the legitimate route (Dr4, Dr3, Dr2, Dr1 and the ground control station) and the route creation rate through the wormhole (Dr4, A1, A2 and the ground station). As we see from Figure 3.45, the creation rate along the genuine route is around 85% for the whole duration of the simulation, whilst the result for the wormhole route is 0%. The route creation rate using SUAP does not add up to 100% as, in this case, node mobility comes into play and creates fluctuations in connections between neighbors. These values should be compared with our results for the AODV protocol, shown in the same figure. Without the detection mechanism, the route establishment rate is around 90% over the wormhole connection. This total is higher than that achieved by SUAP, as the two attackers are stationary and the links along the route are thus subject to less fluctuation. The wormhole attack may thus appear to improve performance depending on the network topology and the selected metric. That being said, without our detection protocol, video traffic is directed toward the wormhole.

Furthermore, the route creation rate over the genuine route using AODV is significantly lower, being close to 0%. This 0% value is never achieved, as the mobility characteristic of node Dr4 may cause it to be out of range of the attacker A2 for a short period of time, during which it communicates with its genuine neighbor Dr3. This exchange is negligible, however, as the route creation rate over the connection (Dr4, Dr3, Dr2, Dr1, GCS) is around 0.18% for the whole period of communication.

These results demonstrate the capacity of our approach to retain a genuine route through active, legitimate nodes, and to ignore routes through a wormhole. We thus obtain a packet delivery rate of around 85%, similar to that obtained for AODV through simulation (88%). We also note that, in the absence of a wormhole detection mechanism, the percentage of established routes passing through the genuine node is markedly lower. Our results therefore show the SUAP protocol to be effective against wormhole attacks, as described above.

3.9.4. *Validation by performance evaluation: discussion*

In this section, we have described the validation of the routing and security partitions of the SUAP protocol. This validation by performance evaluation was carried out, on the one hand, using a tool which combines emulation (virtual machines) with simulation (the OMNET++ tool), and, on the other hand, in a real environment using Delair-Tech DT18 drones. The SUAP protocol was designed using the model-driven methodology presented in Chapter 2. Following the general idea at the heart of this rapid prototyping methodology, we chose to use these experimental tools to retain a single version of the SUAP source code, suitable for execution on both virtual machines and the final embedded systems. Using real drone movement traces and a Linux system similar to that used by the drones, we were able to create practical and realistic scenarios. The results we obtained in validating the routing functions show that our algorithm, developed using a model-driven approach, responds to the different services required by the ad hoc environment. It produces acceptable results in terms of delivery rates, end-to-end delay, overheads, route re-establishment delays and route stability. As such, we may state that the routing partition of our communications architecture responds successfully to the specification given at the outset.

We also carried out simulations and real tests to validate the security mechanisms in the SUAP protocol. Our test architecture was made up of three zones and three ground stations, one of which was a communicating network node. Our test configuration involved a communications system featuring a fleet of cooperating drones, piloted by ground stations. The performance of the SUAP protocol was validated through a set of tests. The results, as given above, indicate that SUAP is able to ensure the security of communications between nodes by providing message authentication services. It is also able to detect the presence of wormholes, and ensures that only genuine nodes can transmit data packets. Furthermore, it respects the time constraints imposed by the use of real-time traffic, with acceptable route re-establishment and end-to-end delays. The SUAP protocol offers more than satisfactory performance in comparison with the AODV protocol presented in the previous chapter. The values obtained here are sufficient to implement a secure communications architecture for a fleet of drones.

Additionally, we highlighted the way in which a model-driven methodology may be used, validating software partitions using performance testing, and ensuring conformity between the specification and the different functions of the embedded software product.

Conclusions and Perspectives

Conclusions and directions for future research

In this book, we have presented a new rapid prototyping method, which is used to design complex embedded systems. This method was initially developed for the field of aeronautics (focusing on drone systems), but has the potential to be adapted for other critical domains. We feel that our method, or a variation of our method, may be put to profitable use for the development and implementation of any embedded software system with critical safety and security requirements.

We have given a detailed description of the operation of our model-driven method, based on high-level artifacts, which both simplifies the definition of a system and facilitates its validation and verification. We notably discussed the way in which the formal methods may be used to accelerate the verification process and to reduce the number of unit tests to be carried out.

A concrete example of the use of our method was presented in Chapter 4. Considering the definition of a network of communicating drones, we were able to illustrate the representational capacities of our method, its capabilities in producing high-performance embedded software code, and its performance in verifying and validating the final system model. In parallel, we supplied additional examples of complementary validation. We presented and applied a hybrid simulation and emulation tool for UAANET networks in validating certain security functions of our ad hoc routing protocol. Finally, we presented the results and contributions of real experiments. These real-world tests provide a valuable complement to those carried out in a hybrid simulation environment. They notably highlight the capacities of our software solution for secure data routing within an ad hoc drone network. Our results demonstrate the utility of our software development method, using a high-level model to develop a high-performance embedded system.

The work presented in this book raises a number of prospects for future development. For instance, our method might be applied in other domains outside of the field of aeronautics; we feel that it could be of particular value for other critical systems, such as rail or automobile communication networks. In terms of our communicating UAANET network, it is important to note that the work on protocols presented here focused on the network layer, with the aim of ensuring safe and secure transfers of data between drones. It would be interesting to consider other solutions to reinforce system security at a higher level (such as the application layer). This scientific contribution would offer valuable perspectives, applicable to ad hoc networks and to any other type of communication network involving one or more drones. Our team is currently working on this subject as a focus for future development.

Prospects

The research presented here may be extended in a number of different directions.

Complementary performance study for validation of the SUAP protocol

The performance studies of different functions of the SUAP protocol presented here are sufficient to validate our work and the functionalities of the protocol. However, the performance assessment phase presented in the previous chapter could be improved in a number of different ways, notably using additional material resources which were not available at the time of writing (additional DT18 drones), and requiring more time than we had available. Potential improvements might include:

1) Increasing the number of nodes used in real experiments. We only used four communicating nodes, but it would be useful to study the behavior of the SUAP protocol with a higher node density. This would also enable us to verify our defense against wormhole attacks in a real environment.

2) Creating a mobile attacker for use in validating the security partition, with the capacity to interact directly with drones in-flight, in order to measure the impact of our authentication mechanism. For example, it would be interesting to place a flying attack node, attempting to exchange fake routing packets, within range of a target node.

3) Validating our proposals for wormhole defense in a real environment, something which we were not able to do due to a lack of hardware. It would be interesting to test this mechanism using a fleet of drones.

Application security for communications in a UAANET environment

In this book, we have focused on the network layer, using authentication and integrity verification mechanisms for signaling traffic in network routing. These mechanisms are designed to ensure the integrity of the routes used to transmit data, but we have not considered the security of data packets themselves. An attacker may be able to obtain this information by illicit listening to the wireless medium (man in the middle attack). In this case, encryption techniques must be used to guarantee confidentiality.

Questions may be raised concerning the relevance of this security layer at the application level in terms of user requirements; however, its necessity is demonstrable. Considering the specification defined in association with industrial players in the context of this book, we see that Delair-Tech drones are often used for the surveillance of industrial zones with significant confidentiality constraints. The data transmitted in these cases must not be accessible to third parties in order to prevent attackers from gaining knowledge of strategic access points within the zone.

The proposals made in existing literature are mostly based on the use of hybrid cryptography, encrypting data as it leaves the emitting node and decrypting it on reception [HAN 11]. However, in the context of our UAANET network, we need to consider the rigid time constraints imposed on application traffic. To our knowledge, this characteristic has not yet been considered in detail. This area offers interesting prospects for future work on identifying an application security solution, ensuring the confidentiality of payload data while respecting the specific requirements of real-time communications.

Designing a key management system for UAANETs

The use of an encryption protocol is essential to ensure secure communications between drones and the ground control station. Digital signature and hash systems, based on encryption procedures, have already been discussed in this work. However, it is important to note that these methods require the use of an encryption key management system in order to be coherent and effective.

To date, no specific proposals have been made in this area for use with UAANET networks; we therefore need to consider developments designed for MANET networks and to study the ways in which they may be adapted for ad hoc drone networks. Certain modifications are needed to respond to the specific characteristics of this type of network. In this work, we considered the creation of a key management infrastructure, presenting the state of the art in this area and analyzing a number of outline proposals. This involved a review of PKIs currently used for ad hoc networks. Generally speaking, classic PKIs are not suitable for use in a MANET network for a number of reasons. First, these networks have no fixed infrastructure. Furthermore, the use of a central certification authority introduces a single point of failure into the system. We also need to take account of the fact that in a mobile network, such as the drone network considered here, nodes are in constant movement and the network topology is thus highly variable. Different types of implementation have been developed to respond to these requirements, and may be divided into four distinct categories: partially distributed PKIs, fully distributed PKIs, certificate chain PKIs and clustering PKIs.

A number of PKI architectures exist, or have been proposed, for the MANET family of networks, all of which share the same aim but present their own specific characteristics. However, none of these proposals correspond precisely to the requirements of the UAANET network. It would be interesting to use existing mechanisms as a basis for developing a specific PKI model for use in this context. In the case of UAANET networks, for example, the PKI system should minimize exchanges between the aircraft and the ground control system. These exchanges are costly, and can increase the delay between the moment a drone enters the network and the moment the GCS is able to establish contact. Furthermore, in certain cases, the GCS remains in its initial position within a UAANET network throughout the entire mission, and this characteristic could be exploited by a PKI infrastructure, considering the GCS as a certification authority. In this case, a drone entering the network would begin by establishing a connection with the GCS in order to prove its authenticity.

Bibliography

[AKR 14] AKRAM R.N., BONNEFOI P.-F., CHAUMETTE S. *et al.*, "Improving security of autonomous UAVs fleets by using new specific embedded secure elements a position paper", *2nd AETOS International Conference on Research Challenges for Future RPAS/UAV Systems*, IEEE, Bordeaux, 2014.

[ALS 11] ALSHABTAT A.I., DONG L., "Low latency routing algorithm for unmanned aerial vehicles ad-hoc networks", vol. 5, no. 8, pp. 629–635, 2011.

[ALV 06] ALVES-FOSS J., OMAN P.W., TAYLOR C. *et al.*, "The MILS architecture for high-assurance embedded systems", *International Journal of Embedded Systems*, vol. 2, nos 3–4, pp. 239–247, 2006.

[ARA 16] ARATHY K., SMINESH C., "A novel approach for detection of single and collaborative black hole attacks in MANET", *Procedia Technology*, vol. 25, pp. 264–271, 2016.

[ARM 05] ARMANDO A., BASIN D., BOICHUT Y. *et al.*, "The AVISPA tool for the automated validation of internet security protocols and applications", *International Conference on Computer Aided Verification*, Springer, pp. 281–285, 2005.

[ARP 96] ARPS 4754, Certification Considerations for Highly-integrated or Complex Aircraft Systems, SAE, Warrendale, 1996.

[BAC 05] BACIC M., "On hardware-in-the-loop simulation", *44th IEEE Conference on Decision and Control, 2005 and 2005 European Control Conference, CDC-ECC'05*, IEEE, pp. 3194–3198, 2005.

[BAG 07] BAGHAI T., HILDEMAN V., Avionics Certification: A Complete Guide to DO-178 (Software), DO-254 (Hardware), Avionics Communications, United States, 2007.

[BAR 16] BARTOLOMEU P., ALAM M., FERREIRA J. *et al.*, "Survey on low power real-time wireless MAC protocols", *Journal of Network and Computer Applications*, vol. 75, pp. 293–316, 2016.

[BEK 13] BEKMEZCI I., SAHINGOZ O.K., TEMEL S., "Flying ad-hoc networks (FANETs): a survey", *Ad Hoc Networks*, vol. 11, no. 3, pp. 1254–1270, 2013.

[BER 00] BERRY G., BOUALI A., FORNARI X. *et al.*, "Esterel: a formal method applied to avionic software development", *Science of Computer Programming*, vol. 36, no. 1, pp. 5–25, 2000.

[BÉR 13] BÉRARD B., BIDOIT M., FINKEL A. *et al.*, *Systems and Software Verification: Model-Checking Techniques and Tools*, Springer Science & Business Media, 2013.

[BLA 03] BLANCHET B., COUSOT P., COUSOT R. *et al.*, "A static analyzer for large safety-critical software", *ACM SIGPLAN Notices*, vol. 38, pp. 196–207, 2003.

[BRA 12] BRAGA R.T., TRINDADE JR O., BRANCO K.R. *et al.*, "Incorporating certification in feature modelling of an unmanned aerial vehicle product line", *Proceedings of the 16th International Software Product Line Conference-Volume 1*, ACM, pp. 249–258, 2012.

[BUC 02] BUCHEGGER S., LE BOUDEC J.-Y., "Nodes bearing grudges: towards routing security, fairness, and robustness in mobile ad hoc networks", *Proceedings of 10th Euromicro Workshop on Parallel, Distributed and Network-based Processing*, IEEE, pp. 403–410, 2002.

[CAM 10] CAMERON S., HAILES S., JULIER S. *et al.*, "SUAAVE: combining aerial robots and wireless networking", *25th Bristol International UAV Systems Conference*, pp. 1–14, 2010.

[CER 01] CERVESATO I., "The Dolev-Yao intruder is the most powerful attacker", *16th Annual Symposium on Logic in Computer Science (LICS)*, vol. 1, pp. 1–16, 2001.

[CHE 13] CHENG X., HUANG X., DU D.-Z., *Ad Hoc Wireless Networking*, vol. 14, Springer Science & Business Media, 2013.

[CLA 03] CLAUSEN T., JACQUET P., Optimized link state routing protocol (OLSR), Report, 2003.

[DAL 08] DALAMAGKIDIS K., VALAVANIS K., PIEGL L., "On unmanned aircraft systems issues, challenges and operational restrictions preventing integration into the National Airspace System", *Progress in Aerospace Sciences*, vol. 44, no. 7, pp. 503–519, 2008.

[DAN 09] DANIEL K., DUSZA B., LEWANDOWSKI A. *et al.*, "AirShield: a system-of-systems MUAV remote sensing architecture for disaster response", *2009 3rd Annual IEEE Systems Conference*, IEEE, pp. 196–200, 2009.

[DIJ 68] DIJKSTRA E.W., "Letters to the editor: go to statement considered harmful", *Communications of the ACM*, vol. 11, no. 3, pp. 147–148, 1968.

[DRO 16] DRONELIFE, "dt18", available at: http://dronelife.com/cms/product/DT-18, 2016, accessed 24 November 2016.

[ELS 09] ELSTON J., FREW E.W., LAWRENCE D. *et al.*, "Net-centric communication and control for a heterogeneous unmanned aircraft system", *Journal of Intelligent and Robotic Systems*, vol. 56, nos 1–2, pp. 199–232, 2009.

[ENG 14] ENGOULOU R.G., BELLAÏCHE M., PIERRE S. *et al.*, "VANET security surveys", *Computer Communications*, vol. 44, pp. 1–13, 2014.

[ERM 08] ERMAN A.T., VAN HOESEL L., HAVINGA P. *et al.*, "Enabling mobility in heterogeneous wireless sensor networks cooperating with UAVs for mission-critical management", *IEEE Wireless Communications*, vol. 15, no. 6, pp. 38–46, 2008.

[EST 13] ESTRADA R.G., SASAKI G., DILLABER E., "Best practices for developing DO-178 compliant software using model-based design", *AIAA Infotech@ Aerospace (I@ A) Conference*, p. 4566, 2013.

[FAD 10] FADLULLAH Z.M., TALEB T., SCHÖLLER M., "Combating against security attacks against mobile ad hoc networks (MANETs)", *Security of Self-Organizing Networks: MANET, WSN, WMN, VANET*, vol. 173, pp. 1–13, 2010.

[FOR 07] FORSMANN J.H., HIROMOTO R.E., SVOBODA J., "A time-slotted on-demand routing protocol for mobile ad hoc unmanned vehicle systems", *Defense and Security Symposium*, International Society for Optics and Photonics, pp. 65611P–65611P, 2007.

[FU 07] FU B., DASILVA L.A., "A mesh in the sky: a routing protocol for airborne networks", *MILCOM 2007-IEEE Military Communications Conference*, IEEE, pp. 1–7, 2007.

[GIL 04] GILBERT H., HANDSCHUH H., "Security analysis of SHA-256 and sisters", *Selected Areas in Cryptography*, Springer, pp. 175–193, 2004.

[GOK 16] GOKHALE V., GHOSH S., GUPTA A., "Classification of attacks on wireless mobile ad hoc networks and vehicular ad hoc networks", *Security of Self-Organizing Networks*, Auerbach, p. 195, 2016.

[GOL 93] GOLDBERG A.V., RADZIK T., "A heuristic improvement of the Bellman-Ford algorithm", *Applied Mathematics Letters*, vol. 6, no. 3, pp. 3–6, 1993.

[GRA 11] GRAF T., "Netlink Library (libnl). 2011", *Dostupné z*, available at: http://www. infradead. org/~tgr/libnl/doc/core.html, 2011.

[GUA 05] GUANG L., ASSI C., "On the resiliency of mobile ad hoc networks to MAC layer misbehavior", *Proceedings of the 2nd ACM International Workshop on Performance Evaluation of Wireless Ad Hoc, Sensor, and Ubiquitous Networks*, ACM, pp. 160–167, 2005.

[GUI 14] GUIDE G.S., Matlab, M.A.T.L.A.B., 7.12. 0 (R2014b), Fuzzy logic Toolbox, 1984–2014, The MathWorks, pp. 1–5, 2014.

[GUP 10] GUPTA A.K., SADAWARTI H., VERMA A.K., "Performance analysis of AODV, DSR & TORA routing protocols", *International Journal of Engineering and Technology*, vol. 2, no. 2, p. 226, 2010.

[GUP 15] GUPTA L., JAIN R., VASZKUN G., "Survey of Important Issues in UAV Communication Networks", IEEE, 2015.

[HAN 11] HANAFY A.A., NOURELDIN S.H., AZER M.A., "Immunizing the SAODV protocol against routing information disclosure", *2011 International Conference for Internet Technology and Secured Transactions (ICITST)*, IEEE, pp. 330–334, 2011.

[HAR 99] HARRIS B., HUNT R., "TCP/IP security threats and attack methods", *Computer Communications*, vol. 22, no. 10, pp. 885–897, 1999.

[HU 03] HU Y.-C., PERRIG A., JOHNSON D.B., "Rushing attacks and defense in wireless ad hoc network routing protocols", *Proceedings of the 2nd ACM Workshop on Wireless Security*, ACM, pp. 30–40, 2003.

[HU 04a] HU L., EVANS D., "Using directional antennas to prevent wormhole attacks", *NDSS*, pp. 241–245, 2004.

[HU 04b] HU Y.-C., PERRIG A., "A survey of secure wireless ad hoc routing", *IEEE Security & Privacy*, vol. 2, no. 3, pp. 28–39, 2004.

[HYL 07] HYLAND M., MULLINS B.E., BALDWIN R.O. *et al.*, "Simulation-based performance evaluation of mobile ad hoc routing protocols in a swarm of unmanned aerial vehicles", *21st International Conference on Advanced Information Networking and Applications Workshops, AINAW'07*, IEEE, vol. 2, pp. 249–256, 2007.

[JAI 09] JAIN M., KANDWAL H., "A survey on complex wormhole attack in wireless ad hoc networks", *Proceedings of International Conference on Advances in Computing, Control, and Telecommunication Technologies (ACT 09)*, pp. 28–29, 2009.

[JAV 12] JAVAID A.Y., SUN W., DEVABHAKTUNI V.K. *et al.*, "Cyber security threat analysis and modeling of an unmanned aerial vehicle system", *2012 IEEE Conference on Technologies for Homeland Security (HST)*, IEEE, pp. 585–590, 2012.

[JOH 01] JOHNSON D.B., MALTZ D.A., BROCH J. *et al.*, "DSR: the dynamic source routing protocol for multi-hop wireless ad hoc networks", *Ad Hoc Networking*, vol. 5, pp. 139–172, 2001.

[JOH 07] JOHNSON D.B., HU Y., MALTZ D.A., The dynamic source routing protocol (DSR) for mobile ad hoc networks for IPv4, Report, 2007.

[KAN 07] KANNHAVONG B., NAKAYAMA H., NEMOTO Y. *et al.*, "A survey of routing attacks in mobile ad hoc networks", *IEEE Wireless Communications*, vol. 14, no. 5, pp. 85–91, 2007.

[KAR 00] KARP B., KUNG H.-T., "GPSR: greedy perimeter stateless routing for wireless networks", *Proceedings of the 6th Annual International Conference on Mobile Computing and Networking*, ACM, pp. 243–254, 2000.

[KAU 14] KAUR R., SINGH P., "Black hole and greyhole attack in wireless mesh network", *American Journal of Engineering Research*, vol. 3, pp. 1–10, 2014.

[KES 08] KESHENG L., JUN Z., TAO Z., "The clustering algorithm of UAV networking in near-space", *ISAPE 2008, 8th International Symposium on Antennas, Propagation and EM Theory*, IEEE, pp. 1550–1553, 2008.

[KHI 12] KHIAVI M.V., JAMALI S., GUDAKAHRIZ S.J., "Performance comparison of AODV, DSDV, DSR and TORA routing protocols in MANETs", *International Research Journal of Applied and Basic Sciences*, vol. 3, no. 7, pp. 1429–1436, 2012.

[KHO 08] KHOKHAR R.H., NGADI M.A., MANDALA S., "A review of current routing attacks in mobile ad hoc networks", *International Journal of Computer Science and Security*, vol. 2, no. 3, pp. 18–29, 2008.

[KIM 12] KIM A., WAMPLER B., GOPPERT J. *et al.*, "Cyber attack vulnerabilities analysis for unmanned aerial vehicles", *Infotech@Aerospace*, pp. 1–30, 2012.

[KOH 96] KOCHER P.C., "Timing attacks on implementations of Diffe-Hellman, RSA, DSS, and other systems", *Annual International Cryptology Conference*, Springer, Berlin, pp. 104–113, 1996.

[KUI 11] KUIPER E., NADJM-TEHRANI S., "Geographical routing with location service in intermittently connected MANETs", *IEEE Transactions on Vehicular Technology*, vol. 60, no. 2, pp. 592–604, 2011.

[KYA 05] KYASANUR P., VAIDYA N.H., "Selfish MAC layer misbehavior in wireless networks", *IEEE Transactions on Mobile Computing*, vol. 4, no. 5, pp. 502–516, 2005.

[LAK 09] LAKEHAL A., PARISSIS I., "Structural coverage criteria for LUSTRE/SCADE programs", *Software Testing, Verification and Reliability*, vol. 19, no. 2, pp. 133–154, 2009.

[LAU 05] LAUER M., "Building embedded linux distributions with bitbake and openembedded", *Proceedings of the Free and Open Source Software Developers European Meeting (FOSDEM)*, pp. 12–20, Brussels, 2005.

[LI 09] LI Q., ZHAO M., WALKER J. *et al.*, "SEAR: a secure efficient ad hoc on demand routing protocol for wireless networks", *Security and Communication Networks*, vol. 2, no. 4, pp. 325–340, 2009.

[LI 12] LI Y., LUO X., "Cross layer optimization for cooperative mobile ad-hoc UAV network", *International Journal of Digital Content Technology and its Applications*, vol. 6, no. 18, p. 367, 2012.

[LIL 99] LILIUS J., PALTOR I.P., "Formalising UML state machines for model checking", *International Conference on the Unified Modeling Language*, Springer, pp. 430–444, 1999.

[LIN 12a] LIN L., SUN Q., LI J. *et al.*, "A novel geographic position mobility oriented routing strategy for UAVs", *Journal of Computational Information Systems*, vol. 8, no. 2, pp. 709–716, 2012.

[LIN 12b] LIN L., SUN Q., WANG S. *et al.*, "A geographic mobility prediction routing protocol for ad hoc UAV network", *Globecom Workshops (GC Wkshps)*, IEEE, pp. 1597–1602, 2012.

[MAC 99] MACKER J., "Mobile ad hoc networking (MANET): Routing protocol performance issues and evaluation considerations", *American Journal of Engineering Research*, vol. 1, no. 5, pp. 1–8, 1999.

[MAH 08] MAHAJAN V., NATU M., SETHI A., "Analysis of wormhole intrusion attacks in MANETS", *Military Communications Conference, MILCOM 2008*, IEEE, pp. 1–7, 2008.

[MAN 14] MANCINI B., "The great reconnaissance: a UAV project in Niger", *GeoInformatics*, vol. 17, no. 6, p. 6, 2014.

[MAN 15] MANCINI B., AUMONT J.-F., "Mapping quickly and with accuracy", *GeoInformatics*, vol. 18, no. 6, p. 6, 2015.

[MAR 09] MARSHALL D., "Unmanned aerial systems and international civil aviation organization regulations", *NDL Review*, vol. 85, p. 693, 2009.

[MAR 11] MARTIN N., AL-MOUSA Y., SHENOY N., "An integrated routing and medium access control framework for surveillance networks of mobile devices", *International Conference on Distributed Computing and Networking*, Springer, pp. 315–327, 2011.

[MAU 01] MAUVE M., WIDMER J., HARTENSTEIN H., "A survey on position-based routing in mobile ad hoc networks", *Network*, IEEE, vol. 15, no. 6, pp. 30–39, 2001.

[MAX 15] MAXA J.-A., ROUDIERE G., LARRIEU N., "Emulation-based performance evaluation of routing protocols for UAANETs", *Communication Technologies for Vehicles*, pp. 227–240, Springer, 2015.

[MAX 16a] MAXA J.-A., MAHMOUD M.S.B., LARRIEU N., "Extended verification of secure UAANET routing protocol", *Digital Avionics Systems Conference (DASC), 2016 IEEE/AIAA 35th*, IEEE, pp. 1–16, 2016.

[MAX 16b] MAXA J.-A., MAHMOUD M.S.B., LARRIEU N., "How to improve routing protocol security in a RPAS swarm?", *4th AETOS International Workshop on "Research Challenges for Future RPAS/UAV Systems"*, 2016.

[MAX 16c] MAXA J.-A., MAHMOUD M.S.B., LARRIEU N., "Joint model-driven design and real experiment-based validation for a secure UAV ad hoc network routing protocol", *ICNS 2016, 2016 Integrated Communications Navigation and Surveillance Conference*, pp. 1E2-1–1E2-16, 2016.

[MAX 17] MAXA J.-A., MAHMOUD M.S.B., LARRIEU N., "Survey on UAANET routing protocols and network security challenges", *Ad Hoc & Sensor Wireless Networks*, pp. 1E2-1–1E2-16, 2017.

[MOH 09] MOHAMMADIZADEH M., MOVAGHAR A., SAFI S.M., "SEAODV: secure efficient AODV routing protocol for MANETs networks", *Proceedings of the 2nd International Conference on Interaction Sciences: Information Technology, Culture and Human*, ACM, pp. 940–944, 2009.

[MOY 13] MOY Y., LEDINOT E., DELSENY H. *et al.*, "Testing or formal verification: Do-178c alternatives and industrial experience", *IEEE Software*, vol. 30, no. 3, pp. 50–57, 2013.

[NDO 11] NDOUNDAM R., SADIE J.K., "Collision-resistant hash function based on composition of functions", *arXiv preprint arXiv:1108.1478*, 2011.

[NIC 09] NICOLESCU G., MOSTERMAN P.J., "Model-based design for embedded systems", CRC Press, 2009.

[NIU 12] NIU Y., GAO D., GAO S. *et al.*, "A robust localization in wireless sensor networks against wormhole attack", *Journal of Networks*, vol. 7, no. 1, pp. 187–194, 2012.

[OGI 02] OGIER R.G., TEMPLIN F.L., BELLUR B. *et al.*, "Topology broadcast based on reverse-path forwarding (TBRPF)", *draft-ietf-manettbrpf-05.txt*, vol. 1, 2002.

[OHA 05] O'HARA B., PETRICK A., *IEEE 802.11 Handbook: A Designer's Companion*, IEEE Standards Association, 2005.

[PAP 03] PAPARAZZI U., available at: http://paparazzi. enac. fr/wiki, 2003.

[PAT 10] PATEL R., BORISANIYA B., PATEL A. *et al.*, "Comparative analysis of formal model checking tools for security protocol verification", *International Conference on Network Security and Applications*, Springer, pp. 152–163, 2010.

[PAT 15] PATEL A., PATEL N., PATEL R., "Defending against wormhole attack in MANET", *2015 Fifth International Conference on Communication Systems and Network Technologies (CSNT)*, IEEE, pp. 674–678, 2015.

[PAU 12] PAUL H., DAS P., "Performance evaluation of MANET routing protocols", *International Journal of Computer Science Issues (IJCSI)*, vol. 9, no. 4, Citeseer, pp. 9–17, 2012.

[PEI 00] PEI G., GERLA M., CHEN T.-W., "Fisheye state routing: a routing scheme for ad hoc wireless networks", *ICC 2000, IEEE International Conference on Communications*, IEEE, vol. 1, pp. 70–74, 2000.

[PER 01] PERKINS C.E., BHAGWAT P., "DSDV routing over a multihop wireless network of mobile computers", in PERKINS C.E. (ed.), *Ad Hoc Networking*, Addison-Wesley Longman Publishing Co., 2001.

[PER 02] PERRIG A., SZEWCZYK R., TYGAR J.D. *et al.*, "SPINS: security protocols for sensor networks", *Wireless Networks*, Springer-Verlag New York, Inc., vol. 8, no. 5, pp. 521–534, 2002.

[PER 03a] PERKINS C., BELDING-ROYER E., DAS S., Ad hoc on-demand distance vector (AODV) routing, Report, 2003.

[PER 03b] PERKINS C., BELDING-ROYER E., DAS S. *et al.*, "RFC 3561-ad hoc on-demand distance vector (AODV) routing", *Internet RFCs*, pp. 1–38, 2003.

[PER 03c] PERRIG A., JOHNSON D.B., "PACKET LEASHES: a defense against wormhole attacks in wireless ad hoc networks", *IEEE INFOCOM*, pp. 1976–1986, 2003.

[PET 11] PETITCOLAS F.A., "Kerckhoffss principle", *Encyclopedia of Cryptography and Security*, pp. 675–675, Springer, 2011.

[PIR 06] PIRRETTI M., ZHU S., VIJAYKRISHNAN N. *et al.*, "The sleep deprivation attack in sensor networks: analysis and methods of defense", *International Journal of Distributed Sensor Networks*, vol. 2, no. 3, pp. 267–287, 2006.

[PON 08] PONS M.-N., "Analyse du cycle de vie: Comment choisir un logiciel", *Techniques de l'ingénieur. Environnement*, vol. 1, no. G6350, 2008.

[POT 12] POTTER B., Complying with DO-178C and DO-331 using Model-Based Design, MathWorks, Inc. 12AEAS-0090, 2012.

[RAY 04] RAYA M., HUBAUX J.-P., AAD I., "DOMINO: a system to detect greedy behavior in IEEE 802.11 hotspots", *Proceedings of the 2nd International Conference on Mobile Systems, Applications, and Services*, ACM, pp. 84–97, 2004.

[RAY 08] RAYANCHU S., MISHRA A., AGRAWAL D. *et al.*, "Diagnosing wireless packet losses in 802.11: separating collision from weak signal", *The 27th Conference on Computer Communications, INFOCOM 2008*, IEEE, pp. 735–743, 2008.

[RIE 13] RIERSON L., *Developing Safety-Critical Software: A Practical Guide for Aviation Software and DO-178C Compliance*, CRC Press, 2013.

[ROS 09] ROSCOE A., BLAIR S., BURT G.M., "Benchmarking and optimisation of Simulink code using Real-Time Workshop and Embedded Coder for inverter and microgrid control applications", *2009 Proceedings of the 44th International Universities Power Engineering Conference (UPEC)*, IEEE, pp. 1–5, 2009.

[ROS 13] ROSATI S., KRUZELECKI K., TRAYNARD L. *et al.*, Speed-aware routing for UAV ad-hoc networks, arXiv:1307.6350, 2013.

[RUM 04] RUMBAUGH J., JACOBSON I., BOOCH G., *The Unified Modeling Language Reference Manual*, Pearson Higher Education, 2004.

[SAH 14] SAHINGOZ O.K., "Networking models in flying ad-hoc networks (FANETs): concepts and challenges", *Journal of Intelligent & Robotic Systems*, vol. 74, nos 1–2, pp. 513–527, 2014.

[SAL 13] SALOMAA A., *Public-Key Cryptography*, Springer Science & Business Media, 2013.

[SEN 10] SEN J., *Reputation- and Trust-Based Systems for Wireless Self-Organizing Networks*, Aurbach Publications, CRC Press, 2010.

[SHA 11] SHAN H., CHENG H.T., ZHUANG W., "Cross-layer cooperative MAC protocol in distributed wireless networks", *IEEE Transactions on Wireless Communications*, vol. 10, no. 8, pp. 2603–2615, 2011.

[SHI 12] SHIRANI R., ST-HILAIRE M., KUNZ T. *et al.*, "Combined reactive-geographic routing for unmanned aeronautical ad-hoc networks", *8th International Wireless Communications and Mobile Computing Conference (IWCMC)*, IEEE, pp. 820–826, 2012.

[SKI 90] SKIENA S., "Dijkstras algorithm", *Implementing Discrete Mathematics: Combinatorics and Graph Theory with Mathematica*, Addison-Wesley, Reading, pp. 225–227, 1990.

[STA 11] STAUB T., GANTENBEIN R., BRAUN T., "VirtualMesh: an emulation framework for wireless mesh and ad hoc networks in OMNeT++", *Simulation*, vol. 87, nos 1–2, pp. 66–81, 2011.

[STE 11] STEINBACH T., KENFACK H.D., KORF F. *et al.*, "An extension of the OMNeT++ INET framework for simulating real-time ethernet with high accuracy", *Proceedings of the 4th International ICST Conference on Simulation Tools and Techniques, ICST (Institute for Computer Sciences, Social-Informatics and Telecommunications Engineering)*, pp. 375–382, 2011.

[SU 08] SU H., ZHANG X., "Cross-layer based opportunistic MAC protocols for QoS provisionings over cognitive radio wireless networks", *IEEE Journal on Selected Areas in Communications*, vol. 26, no. 1, pp. 118–129, 2008.

[TAM 07] TAMILSELVAN L., SANKARANARAYANAN V., "Prevention of blackhole attack in MANET", *The 2nd International Conference on Wireless Broadband and Ultra Wideband Communications (AusWireless 2007)*, IEEE, pp. 21–21, 2007.

[VAR 12] VARET A., LARRIEU N., "Design and development of an embedded aeronautical router with security capabilities", *Integrated Communications, Navigation and Surveillance Conference (ICNS)*, IEEE, pp. E1–1, 2012.

[VAR 13] VARET A., Conception, Mise en œuvre et évaluation d'un routeur embarqué pour l'avionique de nouvelle génération, PhD Thesis, INSA de Toulouse, 2013.

[VAS 14] VASILIEV D.S., MEITIS D.S., ABILOV A., "Simulation-based comparison of AODV, OLSR and HWMP protocols for flying ad hoc networks", *International Conference on Next Generation Wired/Wireless Networking*, Springer, pp. 245–252, 2014.

[WAT 08] WATSON J., "Virtualbox: bits and bytes masquerading as machines", *Linux Journal*, no. 166, p. 1, 2008.

[WU 07] WU B., CHEN J., WU J. *et al.*, "A survey of attacks and countermeasures in mobile ad hoc networks", *Wireless Network Security*, pp. 103–135, Springer, 2007.

[YAN 04] YANG H., LUO H., YE F. *et al.*, "Security in mobile ad hoc networks: challenges and solutions", *IEEE Wireless Communications*, vol. 11, no. 1, pp. 38–47, 2004.

[YAN 10] YANMAZ E., COSTANZO C., BETTSTETTER C. *et al.*, "A discrete stochastic process for coverage analysis of autonomous UAV networks", *2010 IEEE Globecom Workshops*, IEEE, pp. 1777–1782, 2010.

[YAO 13] YAO L., ZHAO Z., YU G., "Detecting wormhole attacks in wireless sensor networks using hop count analysis", *International Journal on Smart Sensing & Intelligent Systems*, vol. 6, no. 1, pp. 325–331, 2013.

[YI 05] YI P., DAI Z., ZHANG S. *et al.*, "A new routing attack in mobile ad hoc networks", *International Journal of Information Technology*, vol. 11, no. 2, pp. 83–94, 2005.

[ZAN 11] ZANG C., ZANG S., "Mobility prediction clustering algorithm for UAV networking", *GLOBECOM Workshops (GC Wkshps)*, IEEE, pp. 1158–1161, 2011.

[ZAP 02a] ZAPATA M.G., "Secure ad hoc on-demand distance vector routing", *ACM SIGMOBILE Mobile Computing and Communications Review*, ACM, vol. 6, no. 3, pp. 106–107, 2002.

[ZAP 02b] ZAPATA M.G., ASOKAN N., "Securing ad hoc routing protocols", *Proceedings of the 1st ACM workshop on Wireless security*, ACM, pp. 1–10, 2002.

[ZAP 08] ZAPATA M.G., "Key management in wireless mesh networks", in ZHANG Y. *et al.* (eds), *Security in Wireless Mesh Networks*, 2008.

[ZHA 13] ZHAI ZHONGQIANG J.D., REN Y., "The application and improvement of temporally ordered routing algorithm in swarm network with unmanned aerial vehicle nodes", *ICWMC 2013, The Ninth International Conference on Wireless and Mobile Communications*, pp. 549–563, 2013.

[ZHE 14] ZHENG Y., WANG Y., LI Z. *et al.*, "A mobility and load aware OLSR routing protocol for UAV mobile ad-hoc networks", *2014 International Conference on Information and Communications Technologies (ICT 2014)*, IET, pp. 1–7, 2014.

[ZHU 03] ZHU J., ROY S., "MAC for dedicated short range communications in intelligent transport system", *IEEE Communications Magazine*, vol. 41, no. 12, pp. 60–67, 2003.

Index

A, B, C

D, E, F

G, H, I, M

N, P, S, U

FSC
www.fsc.org
MIX
Papier | Fördert gute Waldnutzung
FSC® C083411

Kolibri 360 GmbH
Ferdinand-Jühlke-Straße 7
99095 Erfurt, Deutschland
produktsicherheit@kolibri360.de